Signaling and Communication in Plants

Series Editor
František Baluška
Department of Plant Cell Biology, IZMB
University of Bonn
Bonn, Germany

More information about this series at http://www.springer.com/series/8094

Frantisek Baluska • Monica Gagliano •
Guenther Witzany
Editors

Memory and Learning in Plants

Editors
Frantisek Baluska
Department of Plant Cell Biology, IZMB
University of Bonn
Bonn, Germany

Monica Gagliano
School of Animal Biology
University of Western Australia
Crawley, Western Autralia, Australia

Guenther Witzany
Telos- Philosophische Praxis
Buermoos, Salzburg, Austria

ISSN 1867-9048 ISSN 1867-9056 (electronic)
Signaling and Communication in Plants
ISBN 978-3-319-75595-3 ISBN 978-3-319-75596-0 (eBook)
https://doi.org/10.1007/978-3-319-75596-0

Library of Congress Control Number: 2018939066

© Springer International Publishing AG, part of Springer Nature 2018
This work is subject to copyright. All rights are reserved by the Publisher, whether the whole or part of the material is concerned, specifically the rights of translation, reprinting, reuse of illustrations, recitation, broadcasting, reproduction on microfilms or in any other physical way, and transmission or information storage and retrieval, electronic adaptation, computer software, or by similar or dissimilar methodology now known or hereafter developed.
The use of general descriptive names, registered names, trademarks, service marks, etc. in this publication does not imply, even in the absence of a specific statement, that such names are exempt from the relevant protective laws and regulations and therefore free for general use.
The publisher, the authors and the editors are safe to assume that the advice and information in this book are believed to be true and accurate at the date of publication. Neither the publisher nor the authors or the editors give a warranty, express or implied, with respect to the material contained herein or for any errors or omissions that may have been made. The publisher remains neutral with regard to jurisdictional claims in published maps and institutional affiliations.

Printed on acid-free paper

This Springer imprint is published by the registered company Springer International Publishing AG part of Springer Nature.
The registered company address is: Gewerbestrasse 11, 6330 Cham, Switzerland

Preface

Plants are very complex organisms which use a suite of plant-specific sensory systems to monitor all relevant parameters, both abiotic and biotic, to optimise their metabolism, growth, development, morphogenesis and behaviour in their environment. Because plants are sessile organisms, cognitive processes such as learning, memory and decision-making are essential to enable them to cope with various environmental cues and certainly critical to their survival and reproductive success. Despite this, plants have been conventionally studied as stimulus–response mechanical systems devoid of cognition and behaviour. This volume is part of a series of recent books introducing an alternative perspective to this conventional approach on plants. From this perspective, plants are no longer marginalised into a position of lesser status and capacity but recognised for the complex behavioural repertoire they exhibit, which reveals remarkable cognitive competences. From this perspective, plants are indeed *cognitive* organisms because they are able to consolidate the wide range of sensory inputs collected by the perceptual system into a common signal that activates their body to generate plant-specific behavioural responses. By bringing together under one book cover various inputs from academic environments that span from ecophysiology, chemistry, genetics, behavioural ecology and evolutionary biology to psychology and philosophy, this volume attempts to consolidate this rapidly growing and renewed understanding of plants in order to generate a new engagement with the vegetal world as well as instigate fresh research approaches which are open to (re-)evaluate the significance and meaning of concepts such as memory, learning, intelligence and awareness across systems.

It must be said that the idea that plants exhibit complex and flexible behaviours that entail cognitive processes is not new. In fact, Charles Darwin himself had acknowledged the cognitive sensitivity of plants when he proposed that the tip of plant roots acts like the brain of some animals (i.e. the "root-brain" hypothesis). The latest research corroborates the earlier Darwinian insights and demonstrates that plants do behave and most of their behaviours are adaptive, thus ensuring their survival. For example, roots navigate through soil, spanning large distances and actively modifying their adjacent surroundings via biocommunication, i.e. active exudation and modification of the rhizosphere in close cooperation with a large number of microorganisms and allied fungi. One single plant can have up to hundreds of these coordinated root apex-based Darwinian plant-specific brains, which are in a perfect position to perform parallel communication to provide sense of self and agency to the whole plant. This allows plants to store information about their past experiences and thus have an actively generated background of information about the state of the system at a given time, which can be accessed at a later time to compare and evaluate more recent experiences and respond better or faster in the future. Both memory and learning depend on a variety of successful communication processes within the whole organism. Various studies, including those mentioned in some of the chapters in this book, suggest that epigenetic-related mechanisms could play a key role in plant learning and memory. Epigenetic modifications are essential for the response to the environment at both somatic and transgenerational levels. The latter is especially important for the immediate plant survival and for the long-term adaptation to adverse conditions. As suggested in this volume, plants may also have different forms of memory such as sensory, short- and long-term memory. Besides chemical and molecular memory, electric memory in the form of memristors is also emerging as a possibility in plants.

Clearly, plants have neither an animal type of nervous system nor brain, and thus, the mechanisms for coordinating perception and behaviour must be quite different from the mechanisms present in animals. Nevertheless, plants are faced with the same basic challenges as animals and have evolved ways to interpret the information from their environment to solve them. As the editors, we hope the diverse contributions that constitute this volume will inspire readers to ask some of the long-standing questions as well as new ones and, most importantly, pursue their answers.

Bonn, Germany Frantisek Baluska
Crawley, Australia Monica Gagliano
Buermoos, Austria Guenther Witzany

Contents

Memory and Learning as Key Competences of Living Organisms 1
Guenther Witzany

Deweyan Psychology in Plant Intelligence Research: Transforming
Stimulus and Response 17
Ramsey Affifi

General Issues in the Cognitive Analysis of Plant Learning
and Intelligence 35
Charles I. Abramson and Paco Calvo

Plant Cognition and Behavior: From Environmental Awareness
to Synaptic Circuits Navigating Root Apices 51
František Baluška and Stefano Mancuso

Role of Epigenetics in Transgenerational Changes: Genome Stability
in Response to Plant Stress 79
Igor Kovalchuk

Origin of Epigenetic Variation in Plants: Relationship with Genetic
Variation and Potential Contribution to Plant Memory 111
Massimiliano Lauria and Vincenzo Rossi

Plant Accommodation to Their Environment: The Role of Specific
Forms of Memory 131
Michel Thellier, Ulrich Lüttge, Victor Norris, and Camille Ripoll

Memristors and Electrical Memory in Plants 139
Alexander G. Volkov

Towards Systemic View for Plant Learning: Ecophysiological
Perspective 163
Gustavo M. Souza, Gabriel R. A. Toledo, and Gustavo F. R. Saraiva

Mycorrhizal Networks Facilitate Tree Communication, Learning, and Memory 191
Suzanne W. Simard

Inside the Vegetal Mind: On the Cognitive Abilities of Plants 215
Monica Gagliano

Index 221

Memory and Learning as Key Competences of Living Organisms

Guenther Witzany

Abstract Organisms that share the capability of storing information about experiences in the past have an actively generated background resource on which they can compare and evaluate more recent experiences in order to quickly or even better react than in previous situations. This is an essential competence for all reaction and adaptation purposes of living organisms. Such memory/learning skills can be found from akaryotes up to unicellular eukaryotes, fungi, animals and plants, although until recently, it had been mentioned only as a capability of higher animals. With the rise of epigenetics, the context-dependent marking of experiences at both the phenotype and the genotype level is an essential perspective to understand memory and learning in all organisms. Both memory and learning depend on a variety of successful communication processes within the whole organism.

1 Introduction

Memory skills are an essential feature of living organisms in all aspects of life. It serves as a key competence to better react to environmental circumstances, to better adapt and therefore to represent a crucial identity motif in biological selection profiles. Whether such memorized experiences are genetically fixed and heritable or remain epigenetically variable, memory plays crucial roles for the organism. Until the detection of epigenetic markings, memory was investigated in humans and higher animal species as part of the cognitive processes. Now we know the epigenetic markings are present throughout all domains of life, whereas the cognitive capabilities remain as a core feature of higher animals. Otherwise, we would extend anthropomorphic motifs and central nervous system features into nonanimal domains. Similar to brain-specific capabilities that do not represent cognition, but sub-cognitive features such as sensing, monitoring, interpretation (comparison and

G. Witzany (✉)
Telos-Philosophische Praxis, Salzburg, Buermoos, Austria
e-mail: witzany@sbg.at

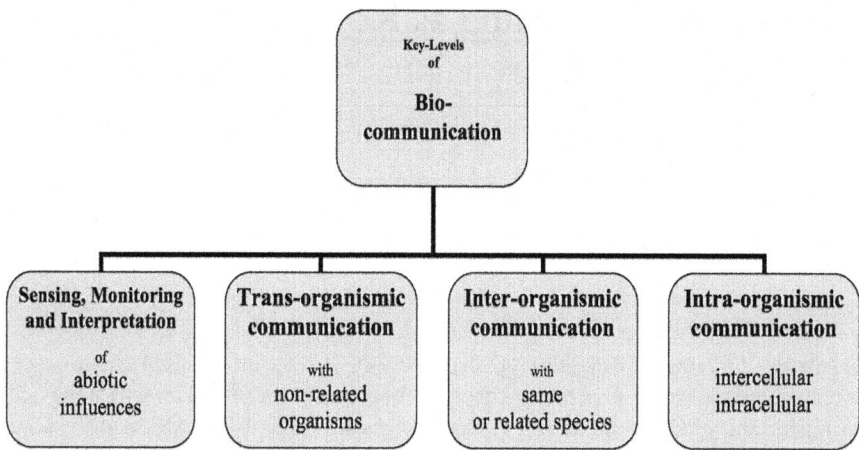

Fig. 1 Memory and learning and its basics, i.e. epigenetic markings, in all organisms of all domains of life depend on complex communicative interactions at several levels of biocommunication

evaluation against stored background information) which can be found in all organisms also communication can be found in all domains of life.

Currently known epigenetic modifications depend on histone modifications—such as acetylation and deacetylation, methylation and demethylation, deimination, phosphorylation and dephosphorylation, isomerization, O-palmitoylation, ubiquitination and ADP-ribosylation—that determine the gene-expression processes. This represents a rich source of tools to mark experienced events of the organism on the genomic level.

Epigenetic markings of certain chromosome sections to target memory relevant modes are essential for different identities of molecule groups, which represent the memorized identity as a kind of "frozen picture" of the total sum of biocommunication processes of an organism in an epigenetically relevant situational context. This means that the epigenetic marking of, for example, extraordinary stress situations—which activate all body parts and their dynamic interactional motifs represented in cells, tissues and organs—takes the "informational content" as the given relevant evaluation for imprinting processes (Fig. 1).

But to evaluate or interpret memory, certain molecular identity groups must play relevant roles within the organism. This means they must trigger a different communication to the interconnected cellular tissues than the previous state where certain memory markings did not exist. If we look at the currently known facts on how organisms store experiences as a memory tool to learn how to better react and quickly adapt within the best energy-saving strategies, we can investigate an abundance of chemicals that serve as signalling molecules for coordination and organization of behavioural patterns. This means that not only memory and learning but all coordination and organization processes in organisms are the result of communicative interactions between cells, tissues and organs (Witzany 2010, 2011, 2012, 2014, 2017; Witzany and Baluška 2012; Witzany and Nowacki 2016). In this respect, the cellular coordination and organization of memory and learning processes are part of a broader realm of the whole communicative interactions in and between cells, tissues and organs of an organism.

If communication processes fail, then coordination and organization such as in memory storing processes by epigenetic imprinting, evaluation by comparison between more recent experiences and the stored background information and changing reaction patterns according to environmental influences within the organism will not occur appropriately. Additionally, this will have consequences for the communication of the individual organism with organisms of the same or related kind or between organisms and non-related organisms.

2 Memory and Learning in All Domains of Life

Why is memory and learning so essential for living organisms in general? And how did these capabilities and techniques of memory and learning—as a result of appropriate interpretation, i.e. comparison of experiences with memorized information and evaluation—emerge?

If we look at evolutionary history, we can identify common patterns that remained the same since the beginning of life, even since the beginning of the RNA world and, later on, cellular life forms. The Darwinian principles of variation and selection are embedded into an unforeseeable environmental dynamics. Such abrupt or long-lasting changes due to climatic, geophysical and gravitational reasons and their interconnections with the environmental life world ecospheres are a characteristic of animated planets. With the advent of RNA world/living organisms, the constant and continued competition for resources needed for survival started as a predominant factor of evolution and adaptation (Atkins et al. 2011).

In addition to abiotic factors that determine evolutionary history in all domains of life, also biotic factors, i.e. behavioural motifs, play essential roles such as competition, cooperation, mating, attack and defence. Besides, we must not forget that each organism is sometimes focused on individual problems within its body such as damage or disease, which may strongly determine its behaviour.

Since the rise of RNA world concepts and the basic knowledge about the roles of viruses and subviral RNAs as genetic parasites and mobile genetic elements (in formatting the gene word order and the roles of non-coding RNAs in the genetic regulation at all stages of cellular processes), we know the importance of group behaviour, group identity and the capability to differentiate between self and non-self to cooperate or to ward off competing biotic agents (Villarreal 2009a, b).

Although memory of experiences seems to be a natural capability of organisms, they usually belong to a complex life world where they communicate with members of the same kind or related organisms. Additionally, the variety of symbiotic and even symbiogenetic interactions demonstrates communicative interactions with non-related organisms. These life-world specific and highly context-dependent interactional patterns are the main resources of memory aspects within the interacting organism.

Flexible genome markings are the precondition of fixing content identity on the genetic level and their regulatory tools, respectively (Slotkin and Martienssen 2007). Certain environmental circumstances, caused by abiotic influences or by biotic agents, may influence the epigenetic storage of such events (Talbert and Henikoff 2014). One best known event is stress situations that cause a different methylation pattern or histone modification, respectively (Santos et al. 2017). But prior to cellular life in the ancient RNA world too, cooperation of RNA stem loops provided the first capability to store information as the starting point of biological memory, by replicating RNA species as recently demonstrated (Urtel et al. 2017).

2.1 *Memory: Context-Dependent Information Storage*

Besides the flexible epigenetic markings that are not part of heritable information transfer, transgenerational immune memory (siRNA, RNAinterference, CRIPRs/Cas) indicates that genetic parasite invasions that are warded off by the immune system will modify and mark those invasive genetic identities to be transferred as memory content via heredity to the offspring.

From the beginning of nucleic acid sequence-based entities on this planet, the behavioural motif of genetic parasites is the driver of constant interactions—whether it be RNA viruses or similar RNA stem loop groups that are in constant interactions with other invading genetic parasites that must be identified, integrated as cooperative parts or warded off (Vaidya et al. 2012). Additionally, this interaction profile means identity problems to the RNA group, because it changes the genetic identity of the RNA group as well as that of the invaded agent (Villarreal and Witzany 2015). This may be disastrous if the former identity was successfully fixed and now may become irrelevant for the host organism, because the function cannot be continued. The new sequence order has to be identified as invasive species and as the relevant target to be warded off (Lambowitz and Zimmerly 2011). On the other side, this flexibility in identity features may cause the rise of a new and unexpected invasive agent identity, being a successful invader of formerly immune hosts (Villarreal 2012). This feature hints at a core feature of life and biotic planets: the constant and continued capability of RNA groups to resist or integrate novel genetic parasites, which drives (i) immune systems, (ii) genetic identities of host organisms and additionally (iii) genetic parasite identities in parallel.

2.2 *Learning and Interpretation*

If an organism in real-life world context with its unique evolutionary and developmental history and identity is able to mark certain genetic setups that represent an environmentally determined specific replication pattern or transcription process, then memory is the result. Memory marks a certain experienced event or multiple similar events to enable this organism to faster and/or more appropriate reaction, if similar situations occur. This capability for better reaction may be termed successful "learning" of the organism based on this stored background information. The organism must differentiate between situations of the same structure without memory and with memory and then be able to evaluate the memory against stored background information. This evaluation process may be termed "interpretation", as stored information leads to "learning", i.e. changing behavioural motifs such as faster/more appropriate reaction to similar real-life experiences. Evaluation of past experiences and comparison with present ones may lead to variable sensing, monitoring, evaluating and making decisions with far-reaching and differentiated consequences. In the long run, biological selection processes will lead to populations who represent an optimized memory/learning/interpretation competence.

3 Memory and Learning in Viruses and Subviral RNA Networks

For a long time, viruses have been considered as molecular invaders unable to replicate themselves. Meanwhile, it is more and more accepted that viruses have an abundance of genes not found in any cellular organism and are therefore older than cellular life. Several researchers have found that viruses, subviral networks and virus-derived parts (such as non-coding RNAs and mobile genetic elements) that are co-opted for host cellular needs play major roles in evolution and development of host organisms (Hayden and Lehman 2006; Smit et al. 2006). Interestingly, short (miRNAs and siRNAs) and long non-coding RNAs and their derivatives, which can function as epigenetic marks of transcriptional gene silencing, also serve as defence tools against transposable elements and viruses (McKeown and Spillane 2014).

Some researchers are of the opinion that the whole genetic content order of cellular organisms is determined by and regulated through such viral and subviral (defective) competencies (Villarreal 2005, 2015). This is because all viruses mark their genomes for self/non-self differentiation, e.g. the virus-first hypothesis suggests that epigenetic markings are transferred to cell-based organisms as infection-derived key competence of viruses that lead to innate and adaptive immune systems in all domains of life, which have been exapted for host purposes (Villarreal 2009a, 2011).

More recently, it was found that even viruses communicate via small peptides and are therefore able to commonly coordinate interactions. More concrete, some phages make decisions whether they should develop into a lytic pathway or choose a temperate (persistent) lifestyle. This communication system strongly influences the decisions of their descendants and may become relevant for epigenetic markings within the phage as well as within the invasion target (Davidson 2017; Erez et al.

2017). All of these features indicate an amplification/suppression system, which is regulated via epigenetic methylation patterns (Villarreal 2009b).

4 Memory and Learning in Akaryotes

Bacteria as well as Archaea, with their different evolutionary histories, are "small but not stupid" (Shapiro 2007). Akaryotes communicate and are therefore able to organize and coordinate their behaviour similar to a multicellular organism (Losick and Kaiser 1997; Schauder and Bassler 2001; Ben Jacob et al. 2004). They continuously monitor and sense their environment and their internal processes such as metabolism, protein regulation, immunity and DNA repair status (Shapiro 2007). Additionally, they are highly competent in cell-cell communication within akaryotic swarm behaviour such as quorum sensing (and quorum quenching) for colonizing biotic or abiotic surface structures and are highly coordinated in complex attack and defence strategies. Also, they have to interact on a transorganismic level with a variety of symbiotic partnerships, being essential for both, such as documented for rhizobacteria within the root zone of plants.

All this coordinated behaviour, as well as biofilm organization that is possible based on signalling within and between akaryotic cells to coordinate and organize, needs some memory system to which actual circumstances can be compared and evaluated as being more relevant to react or represent less priority (Mathis and Ackermann 2016). This leads us to epigenetic imprintings within the akaryotic cell which stores environmental experiences, which are then part of an internal evaluation and also the interpretation system in a simple sense (Casadesús and Low 2006; Oliverio and Katz 2014).

At the origin of such epigenetic imprintings in akaryotes, it is known that these are restriction modification systems, successfully investigated by Kobayashi and his team (Kobayashi 2001; Mruk and Kobayashi 2014). Later on, it was detected that the restriction modification system is a system of counterbalanced, persistent viral infection-derived capabilities, defending the host organism from related genetic parasites. A similar defence system has been found more recently—the CRIPRs/Cas system—which indicates another counterbalanced immune system that is inherently more adaptive and therefore represents an immune function that stores information of viral attacks in a more context-dependent way to ward off genetic parasites that are now memorized more specifically.

5 Memory and Learning in Unicellular Eukaryotes

Unlike the akaryotes, which have highly sophisticated capabilities to communicate, i.e. to generate sign-mediated interactions for various goals and are therefore able to coordinate their single lives within populations like that of a multicellular organism, we can find a more complex behaviour in unicellular eukaryotes also throughout all species. Very good examples can be found in ciliates, where the knowledge about signalling within the ciliate body has been investigated by many research groups (Witzany and Nowacki 2016).

In contrast to akaryotes, in ciliates, the division of labour between the soma and germ line functions is strictly divided, which indicates a different evolutionary ancestor of the two nuclei of ciliates. In unicellular eukaryotes, we are confronted with the evolution of the eukaryotic nucleus and the genetically fixed symbiogenetic integration of various formerly free living akaryotes. The Serial Endosymbiotic Theory (SET) of Lynn Margulis explains the origin of nucleated eukaryotic cells by a merging of archaebacterial and eubacterial cells in anaerobic symbiosis, historically followed by acquisition of mitochondria or plastids (Margulis 1996). In contrast to former evolutionary theories which consider ramification as a driving force of evolution, Margulis initiated a paradigmatic change, bringing merging into the focus of the discussion. But the use of terms like "merging", "fusion", "incorporation" and "amalgamation" is less helpful if we look at the genetic level in symbiogenetic processes (Witzany 2006). If symbiosis leads to symbiogenetic processes, to the development of a new species and thus to the disappearance of the formerly independent individuals, then the result is generative DNA processing, in which genetically different gene pools are combined into one genome. This requires a recombination that assimilates the non-self data set into a "self", converting the external into the internal. This also means the epigenetic markings must be adapted.

Which genome editing competences are able to integrate an endosymbiotic genome in a host genome? Manfred Eigen would ask how we should think about the correct rearrangement of the "molecular syntax" (Witzany 1995). Successful DNA/RNA processing requires numerous, specifically tailored enzyme proteins. In all cases, the DNA/RNA-processing enzyme proteins and also the interacting RNAs together with the epigenetic markings are involved in very precisely conducting these varied DNA-processing steps.

The ciliate epigenetic imprinting is a main source of memory storage and learning to quickly and more appropriately react to changing environmental circumstances (Nowacki and Landweber 2009). In limited nutrition environments, the better reaction modus may lead to the more successful survival strategy. Also, in other unicellular eukaryotes, non-coding RNAs play essential roles in gene regulation and its epigenetic marking. Remnants of former infection events by genetic parasites, such as transposable elements, are exapted, i.e. used and integrated in another function than when previously active. In this respect, we may look at exaptation of the small nuclear and nucleolar RNAs (snRNAs and snoRNAs) to regulate cellular genes and in parallel to mediate transgenerational epigenetic inheritance of essential phenotypic polymorphisms (Singh et al. 2014).

6 Memory and Learning in Fungi

Epigenetics of fungal organisms may be investigated as reversible heritable changes in gene expression without changes in DNA sequence. This refers to gene regulation such as changes in the chromatin structure, although such changes are not necessarily heritable.

Development and growth of fungal organisms depend upon successful communication processes within, and between, cells of fungal organisms. However, on the other side, sign-mediated interactions are necessary to coordinate behaviour with the same, or related, fungal species and with non-related organisms such as bacteria, plants and animals. In order to generate appropriate response behaviour, fungal organisms must be able to sense, memorize, and interpret indices from the abiotic environment as well as from the ecosphere inhabitants and react to them appropriately. However, these communication and interpretation processes can also fail. In such cases, the overall consequences could be disease-causing or even lethal for the fungal organism. Interestingly, certain rules of fungal communication are very similar to those of animals, while others more closely resemble those of plants.

Fungi are heterotrophs. This means they feed by absorbing dissolved molecules. Prior to that, they must, therefore, secrete digestive enzymes into their environment. In contrast to plants, fungi do not photosynthesize. Their mobility occurs by growth, or in the case of spores, they dissipate through air or water. Fungi serve as the main decomposers in ecosphere habitats. As with animals and plants, seasonality is found in fungi as a part of the circadian rhythm (Dunlap and Loros 2004), e.g. light-regulated physiological processes that coordinate the internal fungal clock, which relies on epigenetic markings as the memory system (Aramayo and Selker 2013; Kronholm et al. 2016).

Because fungi seem to represent less complex multicellular eukaryotes (they also have single-celled species), they have to coordinate a rich signalling repertoire within the fungal body, between fungi and related organisms and between fungi and non-fungal organisms. For all of these coordinations, it is essential to produce signalling molecules. The most powerful tool to have access to the rich chemical vocabulary of fungi is epigenetic silencing for regulating the production of semiochemicals. Because fungi can reach tremendous size and life span, it can be expected that the epigenetic memory storage in fungi is rather specialized and will promote life-saving interactions and suppress dangerous or life-damaging causes (Cichewicz 2012).

Especially, symbiotic interactions in the root zone of plants are multilevel communication processes between various plant root cells, mychorizal fungi and rhizobacteria in a highly complicated and dynamic process in which memory, learning, interpretation and the organization and coordination of variable reaction motifs are essential tools for optimal symbiotic interactions. In this respect, memory and learning of participant organisms are involved and co-dependent on each other. This means if one or more communicating patterns or epigenetic disturbances are indicated, this will have far-reaching consequences for all participants of the whole symbiotic interaction.

7 Memory and Learning in Animals

With the evolutionary invention of the animal central nervous system and the brain organ with its neuronal interaction complexity, a new kingdom arose with a really complex intraorganismic communication competence. The linear information processing in animals strengthens the central nervous system's decision making. In contrast to communicative interaction patterns of viruses, akaryotes, protozoa, fungi and plants, animals use vocal and visible signs to communicate.

Accordingly, the epigenetic marking of experiences in memory formation increases in neuronal patterns (Mercer et al. 2008; Sacktor and Hell 2017). Epigenetic imprintings such as DNA methylation, histone modifications and micro-RNA processing with changes in gene expression according to concrete and context-dependent activities are the main source of memory formation patterns (Barlow 2011). The learning process results out of comparison of concrete interactional situations with the background of memorized informations, which may lead to sustained behavioural change. A second point is passing epigenetically memorized information across generations, which means the acquired information that is memorized will be object to inheritable transport (Blaze and Roth 2013).

In contrast to other kingdoms, animals, in most cases, share a mobility which is not found in any other kingdom (Witzany 2014). This makes them vulnerable to a geometric increase of experiences, even for their symbionts or attack enemies. Another interesting aspect is that simple animals, such as *C. elegans* and *Homo sapiens*, share a similar number of protein coding genes, whereas the non-coding regulatory RNA makes the complete difference. Interestingly, the basic memory storing processes in animals (from insects to humans) are comparable within the general concept of cognition (Menzel 2012; Biergans et al. 2016).

8 Memory and Learning in Plants

Plants have often been viewed and studied as machine-like growth automatons. Today, we know that the coordination of development and growth in plants is made possible only by the use of sign(al)s, rather than by pure mechanics. Plants are sessile, highly sensitive organisms that actively compete for environmental resources, both above and below the ground. As do all living organisms, they must assess their surroundings, estimate how much energy they need for goals and then realize the optimum variant. Similarly, they must take measures to control certain environmental resources; as it has been shown that they perceive themselves and can distinguish between self and non-self (Trewavas 2003, 2005), this capability allows them to protect their territory and ward off parasites.

More than 20 different groups of molecules having communicatory functions have currently been identified, and up to 100,000 different substances—known as secondary metabolites—are actively used in the root zone. Such diversity, as it has been proposed, is necessary due to the high abundance of microbes, insects and related or non-related plant roots in this zone and all the interactions made necessary thereby (Bais et al. 2004; Badri et al. 2009). Integration of signalling molecules into coordinated sensing, production and release is highly complex and must be regulated in a precise, timely manner. This is epigenetically imprinted and regulated (Pikaard and Mittelsten Scheid 2014; Matzke et al. 2009, 2015; Birnbaum and Roudier 2017).

More recently, the youngest of all kingdoms is also in focus in research about capabilities which were formerly restricted to higher animals. Although several features of plants such as plant epigenetics, attack and defence mechanisms as well as mating have been observed to change in certain timescales with the change of environmental circumstances such as abiotic stress or new enemies, memory and learning of plants were not the predominant investigation focus (Boyko and Kovalchuk 2011; Baulcombe and Dean 2014; Haak et al. 2017: Rajewski et al. 2017). Especially the epigenetic marking of stress experiences is well documented (Grativol et al. 2012; Gutzat and Mittelsten-Scheid 2012; Lämke and Bäurle 2017).

Some features in heredity demonstrated that plants have the skills to store genetic information on different levels with different evaluation patterns, which indicate some behaviour motifs of interpretation of incoming information and some sort of choice between variable options.

Plants can, for instance, overwrite the genetic code they inherited from their parents and revert to that of their grand- or great-grandparents (Lolle et al. 2005; Pearson 2005; Weigl and Jürgens 2005). This contradicts traditional DNA-textbook conviction that children simply receive combinations of the genes carried by their parents. Now we know that plants are able to replace less appropriate parental code sequences with the regular code possessed by earlier generations—not the inherited parental sequences are translated and transcribed but the backup copy of grand- or great-grandparents. Under normal conditions, the operative genetic make-up stems from the parents. This means, not only a combination of parental genes is inherited but also ancestral genome regulating features in non-coding DNA.

More recent advances in studying memory skills of plants on the genetic level demonstrated that memory is initiated by binding of a transcription factor, leading to essential changes in the chromatin structure and allowing binding of a poised form of RNA polymerase II to promote the rate of future reactivation (D'Urso and Brickner 2017). Communication of plants is not only triggered by chemical compounds for signalling. Most interestingly, more recent experiments demonstrated that plant roots react to sound input also (Gagliano et al. 2017).

8.1 "Communicative Identity" in Plant Behaviour

The capabilities of plants to store information representing experiences and learn by comparing and evaluating the stored experiences with more recent events raise the question of "cognition"-like capabilities of plants, with possible connotations such as "mind" or even "consciousness" of plants. Besides, in humans, the cognition debate is struggling, because scientific descriptions, in many cases, confuse scientific sentences (to describe observations in the realm of theoretical assumptions to reach a common agreement in highly specialized scientific communities) with an imaginary tool that could depict reality.

There is a clear reason why philosophy of consciousness was abandoned and was replaced by the philosophy of language: It was not possible to reach a satisfactory definition of consciousness for many centuries until today, because of inescapable problems of definition. The various consciousness concepts of the last centuries met the pre-assumptions of humans as they defined themselves. But with the success story of neurobiology, it became clear that the term consciousness is an anthropocentric construction to integrate the signalling interactions between several brain tissues in a unifying narrative. If some of the communication brain parts are deformed or damaged, consciousness looks rather strange, and the deficits in the different concepts of consciousness become obvious (Parvizi and Damasio 2001, 2003). For example, the generalizing term consciousness is like how we speak about the national character of, for example, Austria. But, Austria does not exist, except the 8 million people who coordinate and organize the Austrian lifeworld every day by communication processes that even may fail with unexpectable consequences.

Because all quantitative models of communication—which are basically founded on hidden metaphysical assumptions (all-is-one holism or all-is-many atomism, physicalism, mechanicism, e.g. "cellular machinery")—did not function to coherently explain how two interacting biotic agents can reach a common agreement on how to commonly coordinate, the philosophy of consciousness was replaced by the philosophy of language in a long-lasting discourse between 1920 and 1980 (Witzany 1995, 2000). The quantitative methods ignored the results of this discourse, because they could not reflect their own hidden metaphysical assumptions and got stuck in the narratives of subject-object split or similar narratives (body-mind duality), such as sender-receiver (coding-decoding) models.

In the last decade, there arose a discussion about plant "cognition" or plant "consciousness", because of their yet unknown capability for sensing, monitoring, learning, decision making, etc. (Baluška et al. 2006, 2010; Baluška and Mancuso 2009). Because all these capabilities critically depend on successful communication processes within the plant body, I would like to suggest some kind of "communicative identity". This clearly differs from previous assumptions such as mathematical theories of language and communication, e.g. the game theory, but can be investigated as any behavioural coherence of communicative agents that have a historical identity and try to be successful in the main survival strategies, such as commonly reached coordination and organization (intraorganismic, interorganismic and transorganismic).

With "communicative identity", someone can differentiate signal-mediated interactions on every level of every organism of every kingdom. Additionally, it looks at the primacy of context dependence and—most importantly—of group identity (to belong or not to belong, i.e. self/non-self identification competence), and it investigates all semiochemicals in detail and in general. Clearly, "communicative identity" is absent in abiotic matter. No living agents, no sign-mediated interaction (communication) and no rules of sign use are present if water freezes to ice. A clear cut between living and non-living can be drawn here, from RNA groups up to humans.

In contrast to the game-theoretical definition as strictly selfish-orientated behaviour, meanwhile, we can assume that "communicative identity" means any behavioural coherence within the context of an ecosphere-specific real-life world. Because living organisms are no *solus ipse* entities, but, in most cases, interwoven in a social interacting network with same, related or non-related organisms, the communicative behaviour serves not only for selfish but essentially for common goals of participants of an ecosphere habitat with all its symbiotic repercussions.

9 The Emergence of Self/Non-self Differentiation Competence

In symbiotic processes, species-specific communication competence has to be adapted to trans-species communication processes, which means that symbiotic processes depend on adaptation to signalling codes that transgress species borders in most cases. Natural codes function if three levels of rules are followed: syntactic rules determine coherent combination of signals, pragmatic rules determine how code-using agents interact according to changing contexts, and semantic rules determine which meaning/information can be transported with signals.

As all more complex eukaryotes are colonized by symbiotic akaryote settlers that play vital roles for the benefit of their host, they must have adapted to the host ecology. This means they must be able to communicate within their population to coordinate, e.g. population growth, apoptosis, virulence, measurement, decision making, movement and election (quorum sensing), according to group signalling, which does not confuse similar molecules of other (non-self) communities. This seems rather difficult if we imagine, for example, 500 different bacterial species in the human oral cavity (Kohlenbrander et al. 2005). Besides, bacteria are overruled by viral settlers by a magnitude of 10 (Rohwer et al. 2014).

Obligate persistent viral settlers of bacteria (phages) that integrate into host genomes must be competent to integrate without damage of the genetic content order and benefit for host capabilities to adapt to fast changing environmental contexts. As viral settlers are obligate in all living organisms on earth, symbiotic interactions represent multiple code compatibility between viruses, akaryotes and eukaryotic hosts (Diaz-Munos et al. 2017). To guarantee a highly sufficient population-based communication to coordinate appropriate group behaviour, a competence to differentiate self from non-self is necessary.

In our context, this means that besides the epigenetic context, which is an essential tool for memory and learning capabilities, such contextual markings are also relevant for the reaction patterns on non-self organisms, even if they represent another species or even an organismic kingdom. The epigenetic marking of certain reactions or general behavioural patterns may affect the epigenetic markings of the symbiotic cooperation partner also. We may look here at the complexity of globally interwoven interactions between organisms of the same, related or non-related ecosphere habitats.

10 Conclusions

Memory and learning are common features of all organisms throughout all domains of life. Additionally, the organization and coordination of memory and learning within the cells need basic motifs of communication—signal-mediated interactions—that coordinate a limited number of steps of epigenetic imprinting at various ways to fix memory relevant experiences of the organism either on the genetic/ genomic level or even at the phenotypic level that does not remain as a heritable feature, such as most memorized contents in higher animals. Within the realm of the biocommunication perspective, memory and learning complete a broader realm of communication capabilities as basic characteristics of living organisms.

References

Aramayo R, Selker EU (2013) *Neurospora crassa*, a model system for epigenetics research. Cold Spring Harb Perspect Biol 5:a017921

Atkins JF, Gesteland RF, Cech TR (eds) (2011) RNA worlds. From life's origin to diversity in gene regulation. Cold Spring Harbor Laboratory Press, New York

Badri DV, Weir TL, van der Lelie D, Vivanco JM (2009) Rhizosphere chemical dialogues: plant-microbe interactions. Curr Opin Biotechnol 20:642–650

Bais HP, Park SW, Weir TL, Callaway RM, Vivanco JM (2004) How plants communicate using the underground information superhighway. Trends Plant Sci 9:26–32

Baluška F, Mancuso S (2009) Plant neurobiology: from sensory biology, via plant communication, to social plant behavior. Cogn Process 10:3–7

Baluška F, Hlavacka A, Mancuso S, Barlow PW (2006) Neurobiological view of plants and their body plan. In: Baluška F, Mancuso S, Volkmann D (eds) Communication in plants: neuronal aspects of plant life. Springer, New York, pp 19–35

Baluška F, Lev-Yadun S, Mancuso S (2010) Swarm intelligence in plant roots. Trends Ecol Evol 25:682–683

Barlow DP (2011) Genomic imprinting: a mammalian epigenetic discovery model. Annu Rev Genet 45:379–403

Baulcombe DC, Dean C (2014) Epigenetic regulation in plant responses to the environment. Cold Spring Harb Perspect Biol 6:a019471

Ben Jacob E, Becker I, Shapira Y, Levine H (2004) Bacterial linguistic communication and social intelligence. Trends Microbiol 12:366–372

Biergans SD, Claudianos C, Reinhard J, Galizia CG (2016) DNA methylation adjusts the specificity of memories depending on the learning context and promotes relearning in honeybees. Front Mol Neurosci 9:82

Birnbaum KD, Roudier F (2017) Epigenetic memory and cell fate reprogramming in plants. Regeneration 4:15–20

Blaze J, Roth TL (2013) Epigenetic mechanisms in learning and memory. Wiley Interdiscip Rev Cogn Sci 4:105–115

Boyko A, Kovalchuk I (2011) Genome instability and epigenetic modification—heritable responses to environmental stress? Curr Opin Plant Biol 14:260–266

Casadesús J, Low D (2006) Epigenetic gene regulation in the bacterial world. Microbiol Mol Biol Rev 70:830–856

Cichewicz R (2012) Epigenetic regulation of secondary metabolite biosynthetic genes in fungi. In: Witzany G (ed) Biocommunication of fungi. Springer, Dordrecht, pp 57–69

Davidson AR (2017) Virology: phages make a group decision. Nature 541:466–467

Diaz-Munos SL, Sanjuan R, West S (2017) Sociovirology: conflict, cooperation, and communication among viruses. Cell Host Microbe 22:437–441

Dunlap JC, Loros JJ (2004) The neurospora circadian system. J Biol Rhythm 19:414–424

D'Urso A, Brickner JH (2017) Epigenetic transcriptional memory. Curr Genet 63:435–439

Erez Z, Steinberger-Levy I, Shamir M, Doron S, Stokar-Avihail A, Peleg Y, Melamed S, Leavitt A, Savidor A, Albeck S, Amitai G, Sorek R (2017) Communication between viruses guides lysis-lysogeny decisions. Nature 541:488–493

Gagliano M, Grimonprez M, Depczynski M, Renton M (2017) Tuned in: plant roots use sound to locate water. Oecologia 184:151–160

Grativol C, Hemerly AS, Ferreira PC (2012) Genetic and epigenetic regulation of stress responses in natural plant populations. Biochim Biophys Acta 1819:176–185

Gutzat R, Mittelsten Scheid O (2012) Epigenetic responses to stress: triple defense? Curr Opin Plant Biol 15:568–573

Haak DC, Fukao T, Grene R, Hua Z, Ivanov R, Perrella G, Li S (2017) Multilevel regulation of abiotic stress responses in plants. Front Plant Sci 8:1564

Hayden EJ, Lehman N (2006) Self-assembly of a group I intron from inactive oligonucleotide fragments. Chem Biol 13:909–918

Smit S, Yarus M, Knight R (2006) Natural selection is not required to explain universal compositional patterns in rRNA secondary structure categories. RNA 12:1–14

Kobayashi I (2001) Behavior of restriction-modification systems as selfish mobile elements and their impact on genome evolution. Nucleic Acids Res 29:3742–3756

Kohlenbrander PE, Egland PG, Diaz PI, Palmer RJ (2005) Genome-genome interactions: bacterial communities in intitial dental plaque. Trends Microbiol 13:11–15

Kronholm I, Johannesson H, Ketola T (2016) Epigenetic control of phenotypic plasticity in the filamentous fungus *Neurospora crassa*. G3 (Bethesda) 6:4009–4022

Lambowitz AM, Zimmerly S (2011) Group II introns: mobile ribozymes that invade DNA. Cold Spring Harb Perspect Biol 3(8):a003616

Lämke J, Bäurle I (2017) Epigenetic and chromatin-based mechanisms in environmental stress adaptation and stress memory in plants. Genome Biol 18:124

Lolle SJ, Victor JL, Young JM, Pruitt RE (2005) Genome wide non mendelian inheritance of extra genomic information in Arabidopsis. Nature 434:505–509

Losick R, Kaiser D (1997) Why and how bacteria communicate. Sci Am 276:68–73

Margulis L (1996) Archaeal-eubacterial mergers in the origin of Eukarya: phylogenetic classification of life. Proc Natl Acad Sci U S A 93:1071–1076

Mathis R, Ackermann M (2016) Response of single bacterial cells to stress gives rise to complex history dependence at the population level. Proc Natl Acad Sci U S A 113:4224–4229

Matzke MA, Kanno T, Matzke AJ (2015) RNA-directed DNA methylation: the evolution of a complex epigenetic pathway in flowering plants. Annu Rev Plant Biol 66:243–267

Matzke M, Kanno T, Daxinger L, Huettel B, Matzke AJ (2009) RNA-mediated chromatin-based silencing in plants. Curr Opin Cell Biol 21:367–376

McKeown PC, Spillane C (2014) Landscaping plant epigenetics. Methods Mol Biol 1112:1–24

Menzel R (2012) The honey bee as a model for understanding the basis of cognition. Nat Rev Neurosci 13:758–768

Mercer TR, Dinger ME, Mariani J, Kosik KS, Mehler MF, Mattick JS (2008) Noncoding RNAs in long-term memory formation. Neuroscientist 14:434–445

Mruk I, Kobayashi I (2014) To be or not to be: regulation of restriction-modification systems and other toxin-antitoxin systems. Nucleic Acids Res 42:70–86

Nowacki M, Landweber L (2009) Epigenetic inheritance in ciliates. Curr Opin Microbiol 12:638–643

Oliverio AM, Katz LA (2014) The dynamic nature of genomes across the tree of life. Genome Biol Evol 6:482–488

Parvizi J, Damasio A (2001) Consciousness and the brainstem. Cognition 79:135–160

Parvizi J, Damasio AR (2003) Neuroanatomical correlates of brainstem coma. Brain 126:1524–1536

Pearson H (2005) Cress overturns textbook genetics. Nature 434:351–360

Pikaard CS, Mittelsten Scheid O (2014) Epigenetic regulation in plants. Cold Spring Harb Perspect Biol 6:a01931

Rajewski N, Jurga S, Barciszewski J (eds) (2017) Plant epigenetics. Springer, Cham

Rohwer F, Youle M, Maughan H, Hisikawa N (2014) Life in our phage world. Wholon, San Diego

Sacktor TC, Hell JW (2017) The genetics of PKMζ and memory maintenance. Sci Signal 10:505

Santos AP, Ferreira LJ, Oliveira MM (2017) Concerted flexibility of chromatin structure, methylome, and histone modifications along with plant stress responses. Biology 6:3

Schauder S, Bassler BL (2001) The languages of bacteria. Genes Dev 15:1468–1480

Shapiro JA (2007) Bacteria are small but not stupid: cognition, natural genetic engineering and socio-bacteriology. Stud Hist Phil Biol Biomed Sci 38:807–819

Singh DP, Saudemont B, Guglielmi G, Arnaiz O, Goût JF, Prajer M, Potekhin A, Przyboś E, Aubusson-Fleury A, Bhullar S, Bouhouche K, Lhuillier-Akakpo M, Tanty V, Blugeon C, Alberti A, Labadie K, Aury JM, Sperling L, Duharcourt S, Meyer E (2014) Genome-defence small RNAs exapted for epigenetic mating-type inheritance. Nature 509:447–452

Slotkin RK, Martienssen R (2007) Transposable elements and the epigenetic regulation of the genome. Nat Rev Genet 8:272–285

Talbert PB, Henikoff S (2014) Environmental responses mediated by histone variants. Trends Cell Biol 24:642–650

Trewavas A (2003) Aspects of plant intelligence. Ann Bot 92:1–20

Trewavas A (2005) Green plants as intelligent organisms. Trends Plant Sci 10:413–419

Urtel GC, Rind T, Braun D (2017) Reversible switching of cooperating replicators. Phys Rev Lett 118:078102

Vaidya N, Manapat ML, Chen IA, Xulvi-Brunet R, Hayden EJ, Lehman N (2012) Spontaneous network formation among cooperative RNA replicators. Nature 491:72–77

Villarreal LP (2005) Viruses and the evolution of life. ASM Press, Washington, DC

Villarreal LP (2009a) The source of self: genetic parasites and the origin of adaptive immunity. Ann N Y Acad Sci 1178:194–232

Villarreal LP (2009b) Origin of group identity. Viruses, addiction and cooperation. Springer, New York

Villarreal LP (2011) Viral ancestors of antiviral systems. Viruses 3:1933–1958

Villarreal LP (2012) The addiction module as a social force. In: Witzany G (ed) Viruses: essential agents of life. Springer, Dordrecht, pp 107–145

Villarreal LP (2015) Force for ancient and recent life: viral and stem-loop RNA consortia promote life. Ann N Y Acad Sci 1341:25–34

Villarreal LP, Witzany G (2015) When competing viruses unify: evolution, conservation, and plasticity of genetic identities. J Mol Evol 80:305–318

Weigl D, Jürgens G (2005) Hotheaded healer. Nature 434:443

Witzany G (1995) From the "logic of the molecular syntax" to molecular pragmatism. Explanatory deficits in Manfred Eigen's concept of language and communication. Evol Cognit 1:148–168

Witzany G (2000) Life: the communicative structure. BoD, Norderstedt

Witzany G (2006) Serial endosymbiotic theory (SET): the biosemiotic update. Acta Biotheor 54:103–117
Witzany G (2010) Biocommunication and natural genome editing. Springer, Dordrecht
Witzany G (ed) (2011) Biocommunication in soil microorganisms. Springer, Heidelberg
Witzany G (ed) (2012) Biocommunication of fungi. Springer, Dordrecht
Witzany G (ed) (2014) Biocommunication of animals. Springer, Dordrecht
Witzany G (ed) (2017) Biocommunication of archaea. Springer, Dordrecht
Witzany G, Baluška F (eds) (2012) Biocommunication of plants. Springer, Heidelberg
Witzany G, Nowacki M (eds) (2016) Biocommunication of ciliates. Springer, Dordrecht

Deweyan Psychology in Plant Intelligence Research: Transforming Stimulus and Response

Ramsey Affifi

Abstract This chapter argues that plant intelligence needs to be considered as a teleological activity, whereby the plant organizes and coordinates itself and its environment in a goal-based way. If plant biologists follow behaviourist (and cognitivist) psychology in reducing intelligence to mechanisms of the mind, they risk repeating the dementalization of the organism that was inherent in behaviouristic animal theory. I employ Deweyan transactionalism to explore why a new empirical approach to plant intelligence is necessary and what such an approach might look like.

1 Introduction

Newspapers and other media sources are justifiably excited by a recent study by Gagliano et al. (2016) that appears to demonstrate classical conditioning in *Pisum sativum*, through training the plant to anticipate the onset of light following a neutral cue. Felicitations are due in equal measure to these creative and bold scientists and to the plants themselves. While the comparatively more simple learning phenomena "habituation" and "sensitization" have already been documented (early on by Bose (1906, 1912) but more recently by Abramson et al. (2002) and Gagliano et al. (2014), conditioning has not previously been reported in plant behaviour literature (Trewavas 2014). The significance is monumental: associative learning has been used repeatedly to police a distinction between the capacities (and often, therefore, the ethical status) of animals and plants. If further evidence corroborates these findings, can we expect a transformation in our approach to plants?

I am not so sure. The behaviouristic tradition out of which association studies developed relies on a mindless and mechanistic conception of animals. This has long been the motivation of behaviourist approaches to learning. For example, influential

R. Affifi (✉)
Thomson's Land G.05, Institute for Education, Teaching and Leadership (IETL), Moray House School of Education, University of Edinburgh, Edinburgh, UK
e-mail: Ramsey.Affifi@ed.ac.uk

© Springer International Publishing AG, part of Springer Nature 2018
F. Baluska et al. (eds.), *Memory and Learning in Plants*, Signaling and Communication in Plants, https://doi.org/10.1007/978-3-319-75596-0_2

neuropsychologist Donald Hebb begins the book that launched the study of neural networks by stating that the task of understanding behaviour is to reduce mental phenomena "to a mechanical process of cause and effect" (1949, p. xi). It seems that the research project aimed at finding mechanisms of the mind is alive and well in current plant research. For example, in a paper entitled "Green plants as intelligent organisms", Trewavas models plant intelligence through an "adaptive representational network" and asserts that "the challenge is to understand the mechanism" of this network (2005, p. 417)[1]. Brenner et al. (2006) explain that the plant neurobiology research programme is about "discovering the mechanisms of signaling in whole plants, as well as among plants and their neighbours" (p. 413). Gagliano et al. (2016) discuss how associative learning may be a "universal adaptive mechanism" (p. 1) across life. While this research is exciting and important, the assumption that mechanism is sufficient for understanding—or exposing—the mind is questionable. As a mode of explanation, mechanism has a Midas touch: everything it comes across transforms into physicochemical interactions. As a result, a mechanistic reduction of mental phenomena is no longer about intelligence at all. I fear little vindication of the capacities of plants can therefore be achieved through importing these methodologies into the field of plant studies.

In this chapter, I outline an activity-oriented and goal-based view of intelligence, proposing a testable, naturalistic concept that does not reduce it to mechanism. Understanding mechanisms provides inadequate insight into intelligence if we ignore the purposive activities that organize, channel, and employ the cause-effect relations that mechanistic approaches reveal. Intelligence is better thought of as a self-organizing process whereby a teleologically motivated organism restructures and recoordinates the interactions possible between its various interactants. If we study a cross-section of the interactants under a microscope, we miss the system-level activity that contextualizes these mechanisms and which is properly considered to be intelligent. I recognize that many researchers in the field (including those just quoted in reference to mechanism) also frequently use terms like "goal", "intention", "choice", and "purpose," sometimes with and other times without scare quotes. One purpose of this article is to argue that we must get beyond the notion that there are only two options before us. We are not limited to either treat these words as metaphors and colourful embellishments or as implying that "there is a little man ... concealed" (Dennett 1978) behind and directing the teleology of the organism. A teleological view does not imply vitalism, but rather the emergent capacity to goal-set at the system level through feeding back and integrating component interactions in ways that adjust the organism favourably towards needed ends. We need a naturalistic concept of teleology that can be used as a theoretical resource to explore and expand upon our understanding of intelligent behaviour across the biosphere.

In arguing for a teleological conception of intelligence, I will appeal to situated cognition literature, and in particular to its founder, the pragmatist philosopher and

---

[1] However, in other places, Trewavas (2007, 2014) does acknowledge that intelligence is a purposive transaction between an organism and its environment.

psychologist John Dewey, who was a key theorist articulating concepts that critically anticipated insufficiencies of both the behaviourist and cognitivist research traditions. Although John Dewey articulated an action-centred conception of intelligence even before the Watson and Thorndike rose to fame, he had already articulated many of the criticisms fall upon behaviourism (and cognitivism) to this day. Further, his evolutionary view of the relationship between organisms and environments is strikingly modern and resonates with the interactive approaches found in enactivism (Menary 2007; Chemero 2009), niche construction theory (Odling-Smee et al. 2003), and developmental plasticity (West-Eberhard 2003) (which is not surprising as all are facets of a modern theory of the organism that places organism activity at the centre of novelty in both developmental and evolutionary processes).

In my view, behaviouristic research strategies succeed in identifying a necessary but insufficient component of intelligent behaviour as they do not explicitly consider the organism's active role in the constitution of the stimulus to which they are conditioned. The interaction between an organism and its stimuli is not to be considered mechanically but *teleologically*: does (and how does) the organism constitute its stimulus by integrating it (or not) into its activity? And how do such changes manifest the organism's changed purpose? Ridding biology of teleology has long been considered essential for its progress as a science, but intelligence without teleology is unintelligent. An account of intelligence needs to understand the relationship between mechanical and teleological elements of living organisms. Teleology is one of the ways in which "the whole" constrains "the parts" directing them to behave in coordinated ways vastly unlike purposeless systems. It has a sort of "downward causation" insofar as it organizes or constrains interactions between the various mechanisms involved (Noble 2012; Deacon 2012), but it is not reducible to these interactions.

I suggest that studying the teleology of plant activity is an empirical research program and that it leads to very different concepts, empirical methods, and strategies. It also leads to different attitudes and values. The behaviouristic tradition, with its focus on conditioning, is known to be linked to technocratic control both by its advocates[2] (Skinner and Watson) and by its critics (Manicas 2011). Against the counterproductive view that *the ability to be controlled is a sign of intelligence*, a Deweyan perspective proposes instead that intelligence is *the capacity to organize divergent contingent elements into purposive coordination*. This change of focus locates intelligence in the organism's rather than the scientist's activity, and so it restores a sense of autonomy and respect to the concept.

---

[2]For example, Watson (1913) explains that "the theoretical goal [of behaviourism] is the prediction and control of behavior" (p. 507).

2 Dewey's View

In 1896, psychologist and philosopher John Dewey wrote an article in *The Psychological Review* that undercut prevalent assumptions about the relationship between organisms and their environments. Despite being chosen in 1942 as one of the most influential psychology articles of the century (McReynolds 2015), this article remains, as Manicas puts it "another unfulfilled Deweyan program which anticipates fatal criticisms" (2011, p. 14) of contemporary theory and inquiry. It seems to have been superficially incorporated into discussion but substantively ignored. One reason for this might have been its relative complexity compared with behaviouristic accounts of organisms. Indeed, influential behaviourist (and student of Dewey), J.A. Watson once proclaimed: "I never knew what [Dewey] was talking about then, and unfortunately for me, I still don't know" (cited by Fancher 1979, p. 316). This might well be taken as an appropriate epitaph for both the behaviourist and cognitive traditions in psychology. A second reason, suggested in the introduction, is that psychology attempted to rid itself of teleology, on the premise that such as exorcism was intrinsic to scientific inquiry in any field. Dewey objected to this explicitly, and proposed a psychological theory that contextualized mechanisms within goal-based intelligent activity, and as a result saw that such activity was a necessary part of the causal explanation of the mechanisms themselves. In this article, I will focus on how behaviourism ignores at its peril the insights gleaned in this apparently influential article. I do this for three reasons. First, behaviourism is the dominant tradition within comparative psychology and animal behaviour research.[3] Second, behaviourism is the approach employed by most scientists now studying plant intelligence empirically. Third, most of the relevant modern approaches to understanding the organism/environment relation derived from developmental and evolutionary theories take aim at the basic presuppositions of behaviourism and converge (mostly unknowingly[4]) upon insights that Dewey had

---

[3]Cognitivism critiqued and swept over behaviourism in human learning research in the late 1950s, led by Chomsky (1959), Miller (1956), Bruner (1957), and others. While cognitivism rejected the externalist mechanisms of behaviourism, it does so only by replacing them with the equally mechanistic notion of mental programs and operations. It is interesting that this revolution did not happen with nearly as much gusto in animal learning research (though see Allen and Beckoff 1999). As I argue here, unless the dynamic transaction between purposive organism and environment is taken into account, the flexible and ongoing adaptivity of life will continue to veiled by such epistemological constructs.

[4]There are obviously some exceptions to this claim. According to Moorhead (2015), Dewey had a strong though generally unacknowledged influence on cybernetics through Norbert Weiner. Gottlieb, an early pioneer in what would become developmental systems theory, had been inspired during his undergraduate days through reading Dewey's descriptions of bidirectional and co-constitutive relationship (Griffiths and Tabery 2013), which informed his notion of coaction (2002). "Extended mind theory" philosopher Andy Clark (2008) mentions Dewey early in his book but then passes over acknowledging significant overlaps between some of their concepts. Shaun Gallagher (2009) recognizes Dewey's refashioning of established thinking about organism-environment relations as an important formative influence on "situated cognition" theory, but few

articulated over a century ago. For these reasons, Dewey's critique of behaviourist premises is worth another look.

Dewey (and James and Mead) attempted to develop a concept of mind based on evolutionary principles. On the one hand, they opposed a view of mind that understood some separate subjective homunculus causing changes in the material world. On the other, they opposed the mechanistic view which considered physico-chemical reductionistic explanations as satisfactory causes of mental phenomena. As developers of what would be called functionalist psychology,[5] their view instead saw the active organism transacting with its environment as the essential organic unit. Separating causes and effects in linear ways, one way or another, was inadequate because the mental system was seen to arise through the organization and activity of the organism-environment system itself. As such, they also hoped to develop a psychological theory that would also do away with the pesky dualism between mind and matter.

Unfortunately, the promise of such a psychology was overshadowed by the simplistic concepts of behaviourism (which sought to do away with the dualism merely by asserting the irrelevance of one half of the dichotomy). Animal (and now plant) behaviour to this day continues to develop based on this move.[6] As such, it preserves in implicit form many of the dualisms that Dewey and his colleagues were striving to undo. This article argues that an understanding of plant mentality would benefit from approaching the plant as an integrated system and follows recent cognitive scientists who have called for a reappraisal of Dewey and philosophical pragmatism (Menary 2007; Johnson 2007; Chemero 2009; Gallagher 2014). Deweyan psychology is particularly consonant with modern theories of biology such as developmental plasticity (West-Eberhard 2003), niche construction theory (Odling-Smee et al. 2003), and developmental systems theory (DST; Oyama 2000).[7] But to

---

well-known philosophers in this area other than Johnson (2007) draw substantively upon him. A recent conference seeking to reintegrate philosophical pragmatism and neuroscience may point the way to further future engagement. Conference organizer, John Shook optimistically suggested that "[e]xperimental psychology and cognitive science then rediscovered many pragmatist views of brain cognition and learning in the 1980s and 1990s, and the first decade of the twenty-first century accelerated this trend back to pragmatism" (Shook 2011, p. 2).

[5]Early functionalist psychology is not to be conflated with later "functionalist" conceptions of the mind [such as Putnam (1960, 1967) or Fodor (1968)], which are better thought of as cognitivist programs seeking mechanisms in the form of mental programs, patterns, or heuristic devices. The pragmatic conception of functionalism recognized that mind was part of an *embodied* engagement in the world and is therefore much closer to situated cognition theories that have criticized the cognitivist paradigm.

[6]Much of contemporary behaviourism is Spencerian instead of Deweyan in the sense described here: "Spencer conceives organic action as re-action, as a response to external forces which develop in a largely autonomous way and lay down their own law. Dewey's account is based upon two-way connections between organism and environment, and the power of intelligent action to transform the world" (Godfrey-Smith 1996, p. 102).

[7]Both developmental systems theory and niche construction theory explicitly acknowledge Richard Lewontin (1983, Levins and Lewontin 1985) as a key influence in helping them overcome dualisms such as environment and organism and nature and nurture. It has been acknowledged that Lewontin

show how and why this view is important for plant studies, it is necessary to go into some detail as to what Deweyan transactionalism is, and how it differs from behaviourism, before we can highlight specific ways in which it might inform plant intelligence theory and practice.

An evolutionary approach contrasts strongly with a mechanistic approach. Most obvious in this is a conceptual shift articulated in Bredo (1998): whereas mechanism seeks to define lawlike interactions between pre-established and unchanging entities, evolutionary thinking sees the form of interactants changing in response to one another through an ongoing unchanging process (i.e. natural selection). In an evolutionary view, the interactants and their behaviours are co-constituted. Whatever structural or behavioural stabilities exist are not laws but contingent emergent coordinations between the interactants. An organism constitutes a niche, while the niche puts selective pressures upon the organism. The process is dynamic and ongoing, and causality is bidirectional. Evolution is always co-evolution.

3 Constituting the Stimulus Through Response

Dewey begins his account by describing and then critiquing the concept of a "reflex arc". The dominant psychological theories of his day (which persist in much of animal psychology now) had a linear view of behaviour. A behaviour starts when a stimulus is received through the sensory organs and is then processed by the nervous system, and finally a response is produced through the motor system. For Dewey, such a linear view shrouded from sight the fact that an organism acts as a unified whole.[8]

One error lay in the assumptions underlying the distinction between stimulus and response. Dewey argued that stimulus and response were not independent events

---

shares many similarities with Dewey (Godfrey-Smith 1996; Pearce 2014), and the reason for this is not coincidental. Both attempt to naturalize Hegelian dialectics. Levins and Lewontin (1985) did so primarily through developing the significance of Engels' methodologies on understanding biological systems. Dewey attempted to naturalize the Oxford Hegelians in light of the Darwinian revolution. Lewontin's choice of Engels as his chief source instead of Dewey likely reflects his political affiliation with Marxism rather than Deweyan liberalism. Niche construction theory is now a successful research program, and its founders are key architects in the development of the extended synthesis in evolution (see, e.g. Laland et al. 2015). Given that niche construction theorists are themselves working on integrating social and biological evolution under a single framework (which is not provided by Lewontin) (ex. Laland and O'Brien 2011), it may make sense to explore Dewey, whose sociological, ethical, political, and epistemological projects were all tied to his transactional conception of the organism–environment relation.

[8]He also thought that it perpetuated a number of metaphysical dualisms that were the heritage of nonscientific thinking. In this account, the sharp distinction between sensory and motor replicates the dualism between the mind and body, while stimulus and response reproduce the separation of organism and environment. Because scientific thinking is committed to some sort of monism, Dewey felt that psychology of his day had not gone far enough and still had vestigial superstitions carving out its basic conceptions

corresponding to the sensory and motor elements of the organism, respectively. This is because what serves as a stimulus is itself *constituted* by motor activity by the organism and only continues to be a stimulus through the ongoing motor activity that the organism engages in. All directed sensory activity has a motor basis, while all directed motor activity is also sensory. Moreover, the sensory and the motor are coupled at any level of analysis. Sensorimotor coordination is therefore not a temporal process where sense precedes action. It is instead a deep and continuous integration of each. As such, it is not the case in intelligent organisms that "[s]pecialized cells are evolutionary-optimized for effective translation of sensory input into developmental and motoric output" (Baluška and Mancuso 2009).

Behaviouristic approaches lead to successful (read: predictable) empirical results while maintaining a mechanistic interpretation of the phenomena but are limited in the scope of learning and intelligence that can be studied. This partial success is the result of the fact that the integrated activity of the organism has been split into linear processes considered in isolation (Bateson 1979). However, examining linear arcs of a circular process hides that the organism sets the context of the interaction, the full significance and scope which gives meaning to the operant or classical conditioning. Significantly, through such an empirical approach, organisms will only yield convincing responses correlated to "external" stimuli imposed on them for *well-formed* habits and coordinations, i.e. for patterns of behaviour which have *become* more or less mechanized [on mechanism being the effect rather than cause of intelligence, see Dewey (1922)]. This hides the most essential part of intelligent process, which is the purposive act of constituting the habit in the first place. No wonder, then, that behaviourism's father, Watson, had this to say: "[i]n a system of psychology completely worked out, given the response the stimuli can be predicted; given the stimuli the response can be predicted" (1913, p. 514). There is no room in the behaviourist model for contingent decision making, choice, experimentation, context, or purpose. Plant research should avoid repeating this mistake, which amounts to seeking legitimacy through throwing the baby out with the bathwater.[9]

While plants do not have sensory or motor organs, they do have sensory cells and various sorts of movement and/or trophic activities. It has been posited [by Darwin but also by more recent scholars (Baluška et al. 2009)] that the root system is analogous to a brain in terms of how it centralizes and processes information. While this is an exciting possibility, we will run amuck if we assume that the mind is to be found somewhere in that centralization. This would be to repeat the linear conception of mind that Dewey criticized (sensory information reaches the roots, gets processed, and leads to some sort of physical change) and along with it the view that sees organisms as passively adapting to external situations rather than actively selecting, engaging, and modifying them. If plants are intelligent organisms,

---

[9]In particular, plant biologists should be sceptical of simplistic conceptions of behaviour that seek to do away with purposive intelligence. More nuanced positions (that incorporate some elements of a Deweyan transactional perspective) such as Tolman's (1966) may be more useful than Watsonian behaviourism and its variations.

we should not hope to find this through imposing a mechanistic and linear interpretation of stimulus and response onto their actions. We should rather look for evidence that the plant engages in purposeful and unified sensorimotor coordinations that lead it to modify the significance of what they encounter and alter behaviour accordingly. For example, we should look for evidence of *selective attention* (with the plant *constituting* its own sensations through modifying its physical relationship with the objects around it), rather than automated and contextual responses to external "stimuli" and *modified treatment of things around them* (as the significance of those things which it engages change through the plant's interaction with it) in line with what the plant is actively trying to do.

It is interesting that Dewey recognized the importance of this teleological dimension in plants. He wrote: "Even a plant must do something more than adjust itself *to* a fixed environment; it must assert itself *against* its surroundings, subordinating and transforming them into material and nutriment... The environment must be plastic to the ends of the agent," and defined adjustment actively. It "is not outer Even a plant must do something more than adjust itself *to* a fixed environment; it must assert itself *against* its surroundings, subordinating and transforming them into material and nutriment" (1891, p. 115).

4 Organism and Environment as Parts of a Unitary Biological Process

Dewey's critique of the reflex arc dissociated stimulus from sensation and response from motor activity through reuniting each together in "action". But it had a further effect in challenging the sharp divide between organism and environment. This predates but also prefigures conceptions of the organism/environment relation in modern evolutionary biology and complexity theory.

As Bredo (1998) noted, if we assume that organisms and their environments are separate things, there are three types of explanations as to how they might fit one another. The first possibility is that organisms become adapted to their environments. This view is called externalism by Godfrey-Smith (1996), and it is an explanatory strategy which sees the environment as relatively static and as the source of selection pressures that organisms need to adjust to or perish. The second possibility is that the environment is fitted to the organism. In biology, this view is championed by the internalists, who argue in various ways that developmental or morphological constraints determine the main structures of the organism (e.g. Goodwin 1994). According to this view, the environment appears, is encountered, and is restructured in specific ways determined by the organization of the organism. A third view sees organism and environment as two independent entities that interact in a way that partially fits and partially clashes, an ongoing battle (or dance?) characterized by conflicts and resolutions, but without one being fundamentally constituted and reconstituted by the other. According to Bredo, these three approaches are not "evolutionary" because they are not co-evolutionary in a deep sense. They take the form of one or the other entity in the interaction as given, instead of seeing how each is dynamically constituted by the other. Dewey's evolutionary approach is co-evolutionary in this strong sense. The entities are constituted and reconstituted through the relation rather than the relation constituted by the entities. The organism/environment distinction is emergent out of and dependent upon living process. In other words, it is an apparent (though obviously *necessary*[10]) dualism born out of an original unity. This is what distinguishes transactionalism from interactionalism, as the former still assumes independent interactants outside of the relationship. When searching for mentality within life, we should be looking beyond mechanism, a paradigm which sees only the algorithmic consequences of fixed conditions—be it fixed conditions in the organism or in the environment. We should instead be on the lookout for cases where living processes "restructure or redefine" (Bredo 457) organism/environment relationships—and are restructured and redefined in turn. As Bredo puts it, "[v]iewed actively, the 'environment' is whatever is currently aiding or inhibiting one's actions, and since one's activity changes, the relevant 'environment' changes with it" (458). Such a mutual co-constitution is the basis of real novelty, real creativity, and real evolution and not merely the unfolding of programs. Life is not a program, it is an unfolding adaptive process continuously organizing and reorganizing organism/environment transactions.

Such a view may not be admitted in a neo-Darwinian scenario that views adaptation from an externalist view. For example, Dennett (1995) understands evolution as a series of incremental changes that are vetted by a more or less fixed environmental context. But a transactional view may be implied and actual integral to the emerging extended synthesis view (Pigliucci and Müller 2010), which is appreciative of the notion that the organism is underdetermined and capable of genuine novelty, and that the environment undergoes much more rapid changes as a result of the activity of organisms. The underdetermined nature of organic life is explored in great detail in work on developmental plasticity. For example, West-Eberhard (2003) describes evolutionary change as a process whereby a changing environment is responded to creatively by a phenotypically flexible organism, which only later (and sometimes) gets these changes encoded in DNA. On the other hand, the active role of organisms in modifying their environments is the basis of niche construction theory (Odling-Smee et al. 2003). Both of these theoretical perspectives give the organism/environment transaction a contingent and co-determined quality, but both also see life as organizing environmental and genetic resources for its own purposes rather than being blinded push to and fro by various internal or external causal factors. The organism is no longer an epiphenomenon or point of intersection between two vastly different realms of causality. Rather, the organism is itself that which participates in the way in which "nature" and "nurture" matter and interact, constituting and constructing both its internal and external environments. As such,

---

[10]It is "necessary" because without constituting an environment/organism distinction, life would not be adaptive and cease to exist.

Dewey's view of mind as active organism/environment relationship is as relevant as ever. In my view, a model of intelligence and of learning commensurate with these new approaches to understanding organisms should form the basis of an experimental research program that questions the sufficiency of the purely externalist approach that has characterized behaviouristic research.

A certain dynamism separates Dewey's view of mind and these modern theories of the organism on the one hand, from mechanistic views. For Dewey, an organism changes its environment through acting within it, which in turn feeds back as the context for the next action to take place. The calibration of a child reaching for an object is an example. At each point, the child steers her hand slightly right or left as her relationship with the object in her environment changes according to her purposeful reach. It is only because this process is ongoing and continuous that reaching can be a smooth and efficient process, but this also indicates the extent to which organism and environment co-inform one another. Such a description, it should be noted, is also how we might characterize niche construction, the only difference being the scale on which the process is seen to be working. In niche construction, an organism's metabolism, activities, and decisions alter or regulate its environment, which thereby changes selection pressures that it and other organisms face (Odling-Smee et al. 2003; Lewontin 1983). Here we see how Dewey intuited the spirit of evolution much more thoroughly than did many other philosophers of evolution during his time (Spencer being an obvious example).

On the other hand, West-Eberhard's (2003) view of developmental plasticity sees organisms as actively modifying their internal and external relations in order to flourish in a world with contingencies, acting as a whole, but adjusting form, structure, and behaviour in consequence to how it modifies these relations. She is explicit in recognizing the "internal environment", which "includes such factors as gene products, cells or growing tissues of different kinds, body temperature, and so on" (p. 32), as on equal footing with external factors as the context that organisms construct and accommodate to bring about plasticity of structure and behaviour. A West-Eberhardian take on Dewey might therefore take Dewey a step farther (and with interesting consequences for empirical study), recognizing that an organism does not merely coordinate activities with its external environment but with its internal one as well. Gene regulation might thereby be seen as a purpose-directed teleological coordination, where the organism sets the genes up as "stimuli" according to organismic needs. In such a view, development that may normally be invisible and with no clear or immediate effects on the external environment may nonetheless qualify as "niche constructing" activities (see Affifi 2016 for a further exploration of this idea). This turns out to be a particularly important idea when considering plant behaviour, learning, and intelligence.

This view is consonant with some system-theoretic perspectives on the nature of the organism-environment interaction. According to Palmer (2004), early Hungarian systems theorist Angyal (1941) argued that organism and environment were shorthands that actually referred to the relative autonomy and heteronomy of the various interactants within a biological process. But if this is the case, the organism is not bounded by skin or cell wall. There are heteronomous elements (i.e. elements governed by processes that are not integrated with biological organization) occurring within the confines of the body just as there are autonomous elements (those functionally co-ordinated through biological process) occurring outside of it. In this view, the environment might better be considered as the set of factors that impinge upon life but that have not (yet) been incorporated into living process. To go back to Deweyan language, we might say that the environment constitutes those elements which have not yet been incorporated into activity, have not yet been specified as a stimulus, and which do not yet have a response which coordinates them with the organism's purposes and goals. In this way, West-Eberhard's distinction between organism and environment (which could be inner or outer) is more adequately understood as a teleological distinction, separating those elements of the system that are coordinated for some purposive outcome and those which are interfering with coordination (but are potential resources that can become incorporated into it).

For example, an animal seeing a red fruit to which it reaches for the purpose of consuming has seeing, interpreting, reaching for, and eating the red fruit as integrated components of its very process of living. Against a background that consists of countless neutral or debilitating potential stimuli, the red fruit is constituted as actual stimuli by the way in which it is coordinated into an activity. On the other hand, some gene product (or combination of gene products) may disturb the coordination of a purposive act (such as may occur through epigenetic changes causing autoimmune disorders). In this case, the red fruit is part of the functional organic process of the animal, whereas the gene products are not. When viewed in this way, the distinction between organism and environment turns out to be as misleading as the distinction between response and stimulus. Or, as one commentator describes Dewey's position: "Environment and organism are not separate factors but aspects of one function—one process, one coordination, one life, one experience" (Pearce 2014, 764). This is not surprising considering the environment is considered in classical theory as the stimulus and the organism behaviour is considered the response. By realizing that the red fruit as a "red fruit eliciting grabbing to eat" is as much a response as it is a stimulus, we also recognize that it is as much organism as it is environment. And teleologically considered [i.e. in the sense described in our preceding description of Angyal (1941)], it is not the environment at all, insofar as it is well coordinated with activity and therefore an intrinsic part of activity itself. The distinction between organism and environment emerges out of biological activity and is therefore best defined functionally in reference to it.

When observing a human or an animal, it is easy to establish what "the act" that the organism is engaging in is. A specific purpose can be gleaned from which what the organism pays attention to, what it neglects, and how it behaves can all be put into proper context. Plant studies need to explore the question about what the act could mean for the case of plants and to devise observational and experimental strategies accordingly. Do plants have specific acts? Do they have goals which, when reached, are consummated? Do they, at this point turn to other matters, coordinating transactions differently and for new purposes? Is there something like this happening in how plants follow the von Liebig hypothesis (Paris 1992), where plants engage in selective rather than uniform foraging strategies? Can we observe plant "acts" where events are constituted as stimuli and response in relation to a specific goal in organisms that do not necessarily move (or do not move in obvious ways)? This may require examining movement in a very different way (micromovements, trophic growth, or changes in chemistry within different parts of the plant). Or, on the other hand, is the plants' entire development one long act?[11]

An additional reason that Deweyan transactionalism might benefit the current embodied and situated turn in cognitive theory and research is that Dewey did not merely trouble the organism-environment split, but he also troubled the individual-social divide. When describing human-environment transactions, Dewey was painstaking in his observations that the modes of such transactions are also constituted by the relationship between that person and other people (through language and other forms of signification and meaning-making, which become part of the process whereby humans constitute stimuli in coordinated activity). Most enactivist theory is single-organism focused (Gallagher 2009), though Thompson (2007) and Affifi (2015) diverge from this general trend. While these more recent scholars may have the benefit of an additional century of empirical data into the manner in which sensation, thinking, and motor activity are united in action, they lack the integrated vision that Dewey suggested. I think that the social dimension is also crucial in plant intelligence theory and research. A number of studies show intraspecies and interspecies communication between plants and other organisms. If this communication matters, then it is expected that it would inform subsequent transactions through modifying what and how these plants constitute and respond to stimuli. This would involve an additional layer of analysis: in what ways are stimuli and responses constituted by plants, not merely by their teleological activity when considered in isolation, but through the ways in which their goals are shaped by semiotic intra- and interspecies interactions?

5 Some Further Implications

Dewey did, however, recognize that in *certain circumstances* an organism appears to treat certain interactions "as" stimuli and others "as" responses, but he was clear that each is *given* significance by the teleological activity of the organism [i.e. it is a "teleological distinction", (Dewey 1963, p. 260)]. The organism is not merely passively waiting about for external stimuli but is generally engaged in purposeful activity, and the sensorimotor system is selectively attuned to its purposes.[12] When

---

[11] And how might considering plants help us develop our understanding of transactionalism in animals and humans (by forcing us to think of different time scales, of different ways of observing 'action', etc.?).

[12] Learning occurs through the ongoing iterations that this sensorimotor activity affords, which transforms the significance of that which it is engaging. In other words, learning occurs when the organism changes the way it constitutes stimulus and response in its activity. The organism is

things are going well, the organism's activity and its environment are well coordinated, and the situation does not break down into these teleological distinctions. When stimulus does occur, it is actually a *problem* constituted by a goal-oriented organism—to which it is seeking a *solution*. In these cases, the organism does not know what to do. It seeks out a stimulus, searching its environment, in order to seek out an appropriate response. A stimulus only emerges *as something to be sought after* in ambiguous situations. And to understand what the organism cares about and is struggling with, it is important to observe how it constitutes and answers the problems it faces.

An additional area that calls for empirical study is therefore the process whereby an organism constitutes a stimulus out of an ambiguous situation. For Dewey, random movements are not the variation upon which conditioning can act (as Skinner, in his mechanistic application of evolutionary theory would have it). Skinner (ex. 1981) required a source of variation for operant conditioning. Specific behaviours could be neither rewarded nor punished if they did not exist. Appealing to externalist Darwinian logic, he assumed a random variation in behaviour generated by the organism with differential environmental fitness. The increased prevalence or extinction of specific behaviours would then be analogous to the spread or loss of a species. Dewey also acknowledged random movements, which he considered to be movements that are not coordinated into a purposive activity. But he noted that these most often occur when an organism is unsure what to do and is attempting to find a stimulus that will bring about a response (i.e. initiate a purposive act). However, Dewey was explicit that from another point of view, these random movements were not random at all: a lack of stimulus was the stimulus that the organism was constituting and responding to and was responding by searching for a particular stimulus relevant to its goals or needs.[13] Of course, because the sensory and motor are always connected, this is itself still a partial way of explaining the phenomena. If the stimulus is uncertain, then so is the response. As the organism searches for a stimulus, it is also searching for a response, each is equally certain or equally uncertain as the case may be. In Dewey's own words, "[t]he real problem may be equally well stated as either to discover the right stimulus, to constitute the stimulus, or to discover, [or] to constitute, the response" (1963, p. 263). This point can be made concretely through the following example: "consider a dog trying to get a ball on the other side of a fence, for example. It first runs one way, then the other, seeing which approach makes the ball more accessible. Its thinking is conducted through actions that continually alter the situation. Through such action, stimulus and response help redefine each other until a promising whole action can be found"

---

interacting with something in its environment and this process alters the significance of the interaction and thereby the interaction itself. For example, consider Dewey's example of a child reaching towards a flame. At each moment, active sensorimotor coordination alters the quality of the sensory object, transforming from something ambiguous to something specific, and now to a source of pain and discomfort. The meaning of the light has transformed as a result of the interaction, as have the subsequent interactions between the child and that light source.

[13] Hence, this alleged randomness is part of another coordinated act on a metalevel.

(Bredo 1997, p. 12). How might we find such analogical searching in plants? How might we search for searching behaviour?

Do plants purposively seek effective stimuli in ambiguous situations? In other words, do they manifest the uncertainty which would be an indicative precondition for "decision making"? And once found, do they then respond to these stimuli purposively? Given that these two types of acts can be obviously discriminated in humans and animals, it is useful to ask what sorts of analogous correlates we might find in plants. We must consider all of the ways in which plants may be exploring and seeking stimuli so as to initiate coordinated purposive activity. Again, the search should not be limited to externally visually overt behaviour. Are there, for example, phases in which a plant regulates genes in what appears to be a random way and other stages where each gene seems to be more clearly linked to function? Chemistry, receptors, VOC releases and ratios, and intra-plant communication pathways are some of the ways in which we might observe phases of relative dis-coordination and coordination. Experimental plant biology will need to develop techniques to study the live organism in action at the biochemical level because the temporal sequence of biochemical changes is itself a movement and behaviour that can potentially reveal plant mentality.

However, exploring plant intelligence as defined here does not require expensive laboratory devices. If Gagliano et al. (2016) experiment is replicated and associative learning by pea plants triangulated in laboratory settings, we should take these conditioning experiments out to field settings. It is here where we are more likely to see Deweyan psychology in action. Where and how conditioning breaks down (if it breaks down) would provide important insights into the nature of plant intelligence. In complex and shifting environments, how do plants juggle competing demands and even resist the experimenter's goal of training them to a neutral stimulus through developing their own goals? Competing demands foreground teleological coordination. This approach has a successful precedent in classroom learning contexts. Newman et al. (1989) replicated classical experimental protocols in classroom settings in order to reveal the complexities of living situations and the extent to which many behaviourist abstractions are laboratory artefacts.

When the teleological development of the system is cut into parts, its purposefulness disappears, and it is reduced to a system that operates mechanically according to simple relations of cause and effect. For this reason, behaviourist studies, which rely on studying habituation and association in their varied forms through examining how environmental changes cause organismic adaptations, can never adequately solve the question of plant (or animal) intelligence. The organism is seen as a passive system prodded to change rather than an active agency that interacts with its environment selectively in order to effect certain consequences. The behaviourist logic will only ever lead to the conclusion that the organism has "learned" the association because it is programmed to be able to form such associations. Rather than eschewing instinct theory, it merely puts it on a metalevel, asserting that we are mechanically open to the moulding influence of the environment within some preset boundaries or contexts. As such, associative learning appears as a mechanism rather than an element of purposive action.

However, a Deweyan perspective sees mechanism as the outcome rather than as the condition for intelligence, and so it does not seek a "mechanism of the mind". In this way, it is consonant with the view from developmental systems theory that considers "instincts" to be hardened habits rather than congenital programs (ex. Gottlieb 2002). Coordinated actions lead to positive outcomes and so are repeated. They *become* automated for *economic* reasons, so that the organism can explore, scrutinize, and make decisions about other things. In this way, a reflexive mechanical behaviour is the eventual outcome of intelligent process, not something that stands in the way of the possibility of intelligence. The life cycle of a habit begins in something very conscious, uncertain, and experimental and ends in something that flows unthinkingly—and sometimes thereby becomes unresponsive to actual conditions. It is worth considering whether and in what ways plants may also exhibit *habit formation*, the generation of mechanism out of indeterminacy as they age, and experience and re-experience similar things repeatedly in their lives. Do plants take more time to develop coordinated responses to insect herbivory early on? Do they eventually develop "the habit" of a reflexive response? And do they employ this sometimes to the point that they become insensitive to conditions where the habit is no longer coordinating purposeful activity appropriately? Longitudinal studies would be necessary to examine this phenomenon conclusively, but statistically analyses of response types and rates to conditions in differently aged plants in naturalistic conditions would provide helpful indications.

References

Abramson CL, Garrido DJ, Lawson AL, Browne BL, Thomas DG (2002) Bioelectric potentials of *Philodendron cordatum*: a new method for investigation of behavior in plants. Psychol Rep 91:173–185

Affifi R (2015) Generativity in biology. Phenomenol Cogn Sci 14:149–162

Affifi R (2016) The semiosis of "side effects" in transgenic interventions. Biosemiotics 9:345–364

Allen C, Beckoff M (1999) Species of mind: the philosophy and biology of cognitive ethology. MIT Press, Cambridge, MA

Angyal A (1941) Foundations for a science of personality. The Commonwealth Fund, New York

Baluška F, Mancuso S (2009) Plant neurobiology: from sensory biology, via plant communication, to social plant behaviour. Cogn Process 10:3–7

Baluška F, Mancuso S, Volkmann D, Barlow PW (2009) The 'root-brain' hypothesis of Charles and Francis Darwin: revival after more than 125 years. Plant Signal Behav 4:1121–1127

Bateson G (1979) Mind and nature: a necessary unity. Bantam, Toronto

Bose JC (1906) Plant response as a means of physiological investigation. Longmans, Green and Co, London

Bose JC (1912) Researches on the irritability of plants. Longmans, Green and Co, London

Bredo E (1997) The social construction of learning. In: Phye GD (ed) Handbook of academic learning: construction of knowledge. Academic, San Diego, CA, pp 3–45

Bredo E (1998) Evolution, psychology, and John Dewey's critique of the reflex arc concept. Elem Sch J 98:447–466

Brenner ED, Stahlberg R, Mancuso S, Vivanco J, Baluška F, Van Volkenburgh E (2006) Plant neurobiology: an integrated view of plant signalling. Trends Plant Sci 11:413–419

Bruner J (1957) Going beyond the information given. Norton, New York, NY
Chemero A (2009) Radical embodied cognitive science. MIT Press, Cambridge, MA
Chomsky N (1959) Review of verbal behavior, by B.F. Skinner. Language 35:26–57
Clark A (2008) Supersizing the mind: embodiment, action, and cognitive extension. Oxford University Press, Oxford
Deacon TW (2012) Incomplete nature: how mind emerged from matter. WW Norton and Company, New York, NY
Dennett D (1978) Brainstorms: philosophical essays on mind and psychology. Bradford Books, Montgomery, VT
Dennett D (1995) Darwin's dangerous idea. Simon & Schuster, New York
Dewey J (1891) Outlines of a critical theory of ethics. Register Publishing Company, Ann Arbor, MI
Dewey J (1922) Human nature and conduct. Dover, Mineola, NY
Dewey J (1963) In: Ratner J (ed) Philosophy, psychology, and social practice. Capricorn Books, New York
Fancher RE (1979) Pioneers in psychology. W.W. Norton, New York, NY
Fodor J (1968) Psychological explanation. Random House, New York
Gagliano M, Renton M, Depczynski M, Mancuso S (2014) Experience teaches plants to learn faster and forget slower in environments where it matters. Oecologia 175:63–72
Gagliano M, Vyazovskiy VV, Borbely A, Grimonprez M, Depczynski M (2016) Learning by association in plants. Sci Rep 6:38427
Gallagher S (2009) Philosophical antecedents to situated cognition. In: Robbins P, Aydede M (eds) The cambridge handbook of situated cognition. Cambridge University Press, Cambridge, MA, pp 35–51
Gallagher S (2014) Pragmatic interventions into enactive and extended conceptions of cognition. Philos Issues 24:10–126
Godfrey-Smith P (1996) Complexity and the function of mind in nature. Cambridge University Press, Cambridge
Goodwin B (1994) How the leopard changed its spots: the evolution of complexity. Orion Books, London
Gottlieb G (2002) On the epigenetic evolution of species-specific perception: the developmental manifold concept. Cogn Dev 1:1287–1300
Griffiths PE, Tabery J (2013) Developmental systems theory: what does it explain, and how does it explain it? Adv Child Dev Behav 44:65–94
Hebb DO (1949) The organization of behavior: a neuropsychological theory. Wiley, New York, NY
Johnson M (2007) The meaning of the body: aesthetics of human understanding. University of Chicago Press, Chicago, IL
Laland KN, O'Brien MJ (2011) Cultural niche construction: an introduction. Biol Ther 6:191–202
Laland KN, Uller T, Feldman MW, Sterelny K, Müller GB, Moczek A, Jablonka E, Odling-Smee J (2015) The extending evolutionary synthesis: its structure, assumptions and predictions. Proc B Royal Soc 282:1–14
Levins R, Lewontin R (1985) The dialectical biologist. Harvard University Press, Cambridge, MA
Lewontin RC (1983) Gene, organism, and environment. In: Bendall DS (ed) Evolution from molecules to men. Cambridge University Press, Cambridge, pp 273–285
Manicas P (2011) American social science: the irrelevance of pragmatism. Eur J Pragm Am Philos 3:1–23
McReynolds P (2015) The American philosopher: interviews on the meaning of life and truth. Lexington Books, Lanham, MD
Menary R (2007) Cognitive integration: mind and cognition unbounded. Palgrave Macmillan, Basingstoke
Miller GA (1956) The magical number seven, plus or minus two: some limits on our capacity for processing information. Psychol Rev 63:81–97

Moorhead L (2015) Down the rabbit hole: tracking the humanizing effect of John Dewey's pragmatism on Norbert Wiener. IEEE Technol Soc Mag 34:64–71

Newman D, Griffin P, Cole M (1989) The construction zone: working for cognitive change in school. Cambridge University, Cambridge, MA

Noble D (2012) Why integration? Integr Med Res 1:2–4

Odling-Smee JF, Laland KN, Feldman MW (2003) Niche construction: the neglected process in evolution. Princeton University Press, New Jersey

Oyama S (2000) The ontogeny of information: developmental systems and evolution. Duke University, Durham, NC

Palmer DK (2004) On the organism-environment distinction in psychology. Behav Philos 32:317–347

Paris Q (1992) The von Liepig hypothesis. Am J Agric Econ 74:1019–1028

Pearce T (2014) The dialectical biologist, circa 1890: John Dewey and the Oxford Hegelians. J Hist Philos 52:747–778

Pigliucci M, Müller GB (eds) (2010) Evolution: the extended synthesis. MIT Press, Cambridge, MA

Putnam H (1960) Minds and machines. Reprinted in Putnam 1975b, pp 362–385

Putnam H (1967) The nature of mental states. Reprinted in Putnam 1975b, pp 429–440

Shook JR (2011) Conference on neuroscience and pragmatism: productive prospects. Philos Ethics Humanit Med 6(1):14

Skinner BF (1981) Selection by consequences. Science 213:501–504

Thompson E (2007) Mind in life. Harvard University Press, Cambridge, MA

Tolman EC (1966) Behaviour and psychological man. University of California Press, Berkeley, CA

Trewavas A (2005) Green plants as intelligent organisms. Trends Plant Sci 10:413–419

Trewavas A (2007) A brief history of systems biology. Plant Cell 18:2420–2430

Trewavas A (2014) Plant behaviour and intelligence. University of Oxford Press, Oxford

Watson JB (1913) Psychology as the behaviorist views it. In: Madden EH (ed) A Sourcebook in the History of Psychology, pp 507–514

West-Eberhard MJ (2003) Developmental plasticity and evolution. Oxford University Press, New York

General Issues in the Cognitive Analysis of Plant Learning and Intelligence

Charles I. Abramson and Paco Calvo

Abstract In this chapter, we identify issues related to the terms behavior, intelligence, and cognition. We also point out problems with inconsistencies in the definitions of learning phenomena and whether plant intelligence needs to be interpreted in cognitive terms. As an alternative to the cognitive model of plant intelligence, we encourage researchers to consider a model combining the radical behaviorism of B. F. Skinner with the ecological psychology of J. J. Gibson where the focus of both perspectives is on the functional analysis of behavior and the recognition of alternative paths to the emergence of intelligence over the course of natural history.

1 Introduction

Over the past decade, there has been a resurgence of interest in the intelligence of plants. In addition to a book (Trewavas 2014), there have been several recent articles that review various aspects of plant intelligence (Abramson and Chicas-Mosier 2016; Affifi 2013; Baluška and Levin 2016; Calvo et al. 2016; Cvrčková et al. 2009; Gagliano et al. 2014, 2016; Karban et al. 2016; Karpinski and Szechynska-Hebda 2010; Marder 2013; Trewavas 2016). Rather than contribute yet another review of the literature, we will focus on what we believe are some general issues that should be considered by those interested in examining the intriguing possibility of plant intelligence. Many of these issues are seldom discussed in this new literature.

C. I. Abramson (✉)
Laboratory of Comparative Psychology and Behavioral Biology, Oklahoma State University, Stillwater, OK, USA
e-mail: Charles.abramson@okstate.edu

P. Calvo
Minimal Intelligence Laboratory, Department of Philosophy, University of Murcia, Murcia, Spain

EIDYN Research Centre and Institute of Molecular Plant Sciences, University of Edinburgh, Edinburgh, UK

At the outset, we would like to acknowledge that the authors of this chapter are a psychologist (CIA) and a philosopher of cognitive science (PC), respectively, and not plant biologists. The issues we identify are approached therefore from a psychological perspective. As the plant intelligence literature extensively borrows terms and perspectives that were once the sole purview of psychologists (e.g., behavior, cognition, perception, learning, memory), we feel that a chapter such as ours can help researchers avoid some of the potential pitfalls that could otherwise cloud the interpretation of plant intelligence. One of us is a comparative psychologist and behaviorist (CIA), while the other is a philosopher of cognitive science with an interest in ecological psychology (PC). The authors of this chapter approach the possibility of plant learning and intelligence from the standpoints of behaviorism and ecological psychology, respectively.

Nevertheless, there are several areas in which they agree. These are:

1. What is behavior?
2. What is intelligence?
3. What is cognition?
4. What is the most promising framework to study plant intelligence?
5. Are there inconsistencies in definitions of learning phenomena?
6. Are enrichment theories needed to study plant learning ecologically?
7. Is there a need to interpret plant intelligence in cognitivist terms?

We do not propose solutions to these issues. However unsatisfactory this may be, we believe that it is important enough to point out these problems to researchers in the hope that a special issue of the journal *Plant Signaling & Behavior* published by the Society of Plant Signaling and Behavior or a workshop during one of the society meetings can be developed in an effort to come to a consensus. We fear that without such a consensus there is a real possibility that an interdisciplinary approach to plant intelligence will not happen and the results that initially showed such promise will continue to be open to a variety of criticism (Alpi et al. 2007; Firn 2004; but see Brenner et al. 2007; Trewavas 2004, 2007) and be relegated to the popular press and public imagination rather than serious interdisciplinary scientific discourse.

2 What Is Behavior?

One of the striking characteristics of the study of behavior is the lack of a consistent definition of what behavior is. *How can one accurately study the behavior of plants if one has no consistent definition of what it is?* In a study examining whether the glossaries of introductory textbooks in psychology, animal behavior, and biology contain a definition of behavior, only 38 textbooks out of 138 texts sampled contained a definition. Moreover, when students in these introductory courses were asked to define behavior, their definitions depended upon the field (Abramson and Place 2005). There were no consistencies across disciplines.

In a recent article that should be widely read by those interested in plant behavior, Cvrčková et al. (2016) were surprised to find that in biology there is also no consistent definition. They cite numerous studies such as Levitis et al. (2009) that found 25 different definitions of behavior. Interestingly, the title of the Levitis et al. (2009) paper is "Behavioural biologists don't agree on what constitutes behaviour." Cvrčková et al. (2016) go on to suggest that an analysis of plant behavior can contribute to a reconceptualization of how to define behavior. We agree, but to do so requires an acknowledgment that this is an issue worth addressing.

3   What Is Intelligence?

As with the definition of behavior, there is no consistent definition of intelligence. In an article aptly titled "A collection of definitions of intelligence," Legg and Hutter (2007) discovered over 70 different definitions. In an earlier investigation, Sternberg and Detterman (1986) queried 24 theorists to define intelligence and received 12 different responses. *How can one accurately study the intelligence of plants if one has no consistent definition of what it is?* Frankly, the answer is you cannot. Even a cursory knowledge of the human intelligence literature recognizes the many problems in measuring and interpreting intelligence (Abramson and Lack 2014). Even the very term is under attack. Schlinger (2003) points out many of the problems with the term. These include no consistent definitions, definitions that change over time, circular reasoning, and errors of reification. In our view using the term intelligence to describe aspects of a plant's behavior without understanding the considerable issues related to the term intelligence is like being tied to a corpse. Why use the term?

As a result of this lack of consistency, researchers can now study naturalistic intelligence, musical intelligence, emotional intelligence, interpersonal intelligence, spatial intelligence, analytical intelligence, creative intelligence, and practical intelligence among others (Gardner 2006; Sternberg 1984). Can we expect to see a resurgence in the study of "musical intelligence" in plants similar to the experiments of Retallack (1973)?

4   What Is Cognition?

Moore notes (2013a, b) that cognitive psychology represents a group of theoretical positions that incorporate mentalism. The trend of studying cognitive processes in plants (and animals) is unfortunate, as it approaches the study of plant (and animal) behavior in disembodied Cartesian terms. More importantly, as with the definition of behavior, there is no consensus across researchers on what cognition is. *How can one accurately study the cognition of plants if one has no consistent definition of what it is?* The answer is you cannot. What is cognition? No one seems to know. A recent study surveying introductory psychology and cognitive textbooks found no consistent definitions. Not only were there no consistent definitions, but similar to the term intelligence, there was an increase in the types of cognition. Cognition now includes analytical cognition, cultural cognition, and holistic cognition (Abramson 2013). If this trend goes unchecked, one can easily imagine studies examining "unconscious cognition" in plants. Information about the use of the term cognition and its history can be found in Whissell et al. (2013) and Chaney (2013), respectively.

Definitions of cognition are so broad that it encompasses almost any behavior. Readers may be surprised to learn that there is no criterion to evaluate whether a behavior can be considered cognitive (Adams 2010; Adams and Aizawa 2008). If there are no objective criteria to establish whether a behavior is "cognitive," how can a plant researcher claim that cognition exists in plants (Aizawa 2014)? The answer is they cannot. As Amsel (1989) noted when asked to define cognition, the founding editor of the journal *Cognitive Psychology* replied that it is "What I like." This sort of reply is typical and has led to harsh criticism toward psychology as a science. A recent study published in *Science* attempted to replicate 100 studies in the areas of cognitive and social psychology. The results were disappointing if not alarming. Of 100 published experiments appearing in three highly ranked journals, 65%, including 50% of the cognitive experiments, could not be replicated, and of those that were replicated, many had reduced effect sizes (Open Science Collaboration 2015). A major problem hindering replication was that many of the terms used by psychologists are not clearly defined, and it is precisely these terms that have been uncritically accepted by many in the plant intelligence community. In discussing definitional issues related to cognitive science, Cvrčková et al. (2016) note that "Similarly, cognitive sciences apparently can live without clear-cut formal definitions of cognition." The work reported by the Open Science Collaboration Project suggests that cognitive science cannot live without such formal definitions.

It has also been suggested that the cognitive perspective constricts research and imposes an almost supernatural position on mental events (Overskeid 2008). This concern is supported by Cromwell and Panksepp (2011) who further emphasize how the overuse and misuse of the term "cognition" is slowing progress in behavioral neuroscience. In addition, the "cognitive" revolution has created a generation of students, and professors, who have little formal knowledge about traditional learning methodologies, proper control procedures, and alternative viewpoints (Abramson 2013).

In reading the plant intelligence literature, there seems to be an uncritical acceptance of interpreting such behavior in terms of cognition – however defined. Why would a plant researcher jump into such a quagmire? We would advocate that plant researchers at least look at an alternative perspective that does away with mentalistic constructs altogether, namely, the one provided by radical behaviorism and ecological psychology in partnership.

5 What Is the Most Promising Framework to Study Plant Intelligence?

It is important to note at the outset that no single theoretical framework can deliver the goods all by itself. Different approaches prove useful for different purposes. In our view, the natural science and history of behavior that behaviorism and ecological psychology, respectively, put forward not only complement each other (Morris 2009) but furnish us with the best way to deal with the study of plant adaptive behavior. Plant scientists somewhat familiar with the psychological literature might be surprised by the alliance that we propose for their consideration. But this is probably due to the two schools having been traditionally portrayed as antithetical or as dealing with different areas of discourse. But both behaviorism and ecological psychology come in many shapes and forms. Putting together the views of, say, either Pavlov, Watson, Hull, or Skinner with those of either Gibson, Baker, Neisser, Costall, or Lee will certainly produce different, more or less complementary outcomes. Some behaviorists (e.g., Skinner) see behavior as holistic, contextual, and molar (Morris 2009), whereas others do not. Some ecological psychologists are partly cognitivists (e.g., Neisser), whereas others are not. Our dance partners are B. F. Skinner and J. J. Gibson. Skinner was contemporary with Gibson and, despite conventional wisdom, shares more than the year of birth (1904, in case our readers are curious). Starting with a molar view of behavior, both behaviorism and ecological psychology ultimately focus on what organisms do and not about muscle contraction and receptor excitation. More generally, they both approached the study of behavior non-mentalistically, complementing and coinciding with each other, however unintentionally, in a number of ways. Skinner and Gibson are equally aware that the environment is part and parcel of behavior and devoted a great deal of effort to describing the environment that surrounds the organism (Gibson 1979; Skinner 1953, 1990). Likewise, neither Skinnerian behaviorism nor Gibsonian ecological psychology posited mediating information-processing states or considered the neurological substrate to be relevant. For obvious reasons, the nonreductionist study of plant intelligence (Calvo 2016; Calvo Garzón and Keijzer 2011) can benefit from endorsing their theoretical stance.

Moreover, a non-mentalistic study of plant intelligence takes us back to the evolutionary theory of Charles Darwin (Costall 2004; Morris 2009) and his pioneering research on plants. Considering Darwin's groundbreaking work, it is difficult to make sense of the fact that cognitivism has continued to ignore the perceptual and behavioral capacities of plants over a century later. Of course, the plant intelligence community is well aware of the importance of Darwin to the overall success of their enterprise (Baluška and Mancuso 2009a, b; Barlow 2006; Kutschera and Niklas 2009) and will surely find rewarding the fact that both Skinnerian behaviorism and Gibsonian ecological psychology can be traced back to Darwin. If Darwin (1859) approached the environment-organism interaction at the biological level, both Skinner and Gibson did the equivalent at the psychological level of interaction. Darwin (1875; Darwin and Darwin 1880) stressed that plants act adaptively, and we can therefore ask what the biological significance of the way they interact with their local environment is. The reciprocal coordination of organism and environment is nonnegotiable to the ecological approach of Gibson (1979), and the same goes for Skinnerian behaviorism when engagement is emphasized (Skinner 1938). It constitutes the proper unit of analysis (Richardson et al. 2008). We may say that both Skinnerian behaviorism and Gibsonian ecological psychology rely on a *relational* ontology whose respective terms are, for example, "operant" and "reinforcer" and "affordance," respectively (Costall 2004), and that involves increasingly sophisticated causal patterns of relations (García Rodríguez and Calvo Garzón 2010).

As originally conceived, operant behavior is a "psychology of action." An operant is goal directed that is affected by contingencies of reinforcement. What is generally not acknowledged is that the organism not only is affected by the contingencies but also must *actively* participate in those contingencies (Lee 1988). Reinforcement is a special type of contingency where behavior is either increased or maintained by the contingency.

In turn, according to Gibson, the affordances of the environment are "what it offers the animal, what it provides or furnishes, either for good or ill. The verb to afford is found in the dictionary, but the noun affordance is not. I have made it up. I mean by it something that refers to both the environment and the animal in a way that no existing term does. It implies the complementarity of the animal and the environment" (1979, p. 127). Affordances are opportunities for behavioral interaction, properties of the surroundings that permit organisms to interact in ways that are relevant to them.

Costall (1984) equates Gibsonian affordances with the Skinnerian discriminative function of the environment for related operant behavior. The connection here goes beyond the "radical" version of Skinner. There is, for example, a parallel between Gibsonian affordances and the "purposive behaviorism" of Tolman (Costall 2004). Tolman (1932) stressed the relational character of behaviors. If an organism requires an environment, behaviors require supports. Of course, Tolman was thinking of rats and experimental devices (do not expect a rat to press a bar or run down an alley in the absence of constraining actual bars, floor, and walls), but the relational tandem "behavior-support" applies equally to plants. Do not expect a vine to climb in the absence of a support suitable for climbing. The climbing plant and the support constitute a coupled system. In Gibsonian jargon, we may say that the vine perceives the possibility to interact with a support that affords climbing (Calvo 2016; Calvo et al. 2014; Carello et al. 2012). In our view, the relational ontologies, reciprocal and functional approaches, and understanding of behavior in holistic, contextual, and molar terms of Skinner and Gibson furnish us with the most promising framework to study plant intelligence. The next two sections deal with behavioral and ecological issues, respectively.

6 Are There Inconsistencies in Definitions of Learning Phenomena?

Many interested in studying the intelligence of plants using behaviorist learning paradigms may be surprised to learn that even here there are inconsistencies (Abramson 1994, 1997). For example, Pavlovian conditioning is often discussed as if there was one standard procedure. This is simply not true. Gormezano and Kehoe (1975) discuss four ways of presenting a conditioned (CS) and unconditioned stimulus (US) that differ based on the how the CS and US are presented. In addition to classical conditioning, there is also alpha conditioning in which the CS elicits a response albeit to a lesser degree similar to the response elicited by the unconditioned stimulus. Alpha conditioning might be better characterized as US-US conditioning. Some researchers consider alpha conditioning not as representing a form of classical conditioning but as instrumental conditioning (Razran 1971). It is important to note that the various procedures used to generate classical conditioning do not measure the same process and that it would be incorrect to consider all of the procedures used to study classical conditioning as measuring the same phenomenon.

In addition to issues related to what is classical conditioning, there are issues related to signaled avoidance and punishment. Avoidance techniques, for example, have been used in the animal behavior literature as a tool to tease apart classical conditioning from instrumental and operant behavior. In avoidance conditioning a response to a cue leads to the omission or postponement of an aversive event. Alternatively in a punishment procedure, a response to an aversive event leads to a decrease in the response.

The definitional issues associated with classical conditioning, signaled avoidance, and punishment also find their way into instrumental and operant behavior. Most behavioral scientists consider instrumental and operant behavior to be synonymous in that both represent behavior controlled by its consequence. However, comparative analysis has revealed that they are not the same (Abramson 1994, 1997). Operant conditioning is more advanced than instrumental conditioning in that it creates "arbitrary" behavior. This distinction is important for those interested in plant intelligence as a plant may indeed exhibit behavior controlled by its consequences (Gagliano et al. 2016), but is this behavior arbitrary? Consider a simple runway experiment where rats can easily be trained to run faster or slower depending on the reinforcement contingencies or an operant situation where rats are trained to push a lever left, right, up, down, or with various degrees of force. No invertebrate learning study yet has conclusively demonstrated that an invertebrate can be trained to run faster or slower depending upon the reinforcement contingencies although they can certainly respond to consequences. If a plant does indeed exhibit operant behavior, paradigms must be developed in which a plant manipulates its environment in response to contingencies of reinforcement. In other words, a plant must show not only that it can manipulate its environment but show you it can use the environment (Abramson 1997).

As mentioned previously (Abramson and Chicas-Mosier 2016), one way to address the problem is through the use of behavioral taxonomies. To our knowledge no plant intelligence study has ever attempted to link their procedures to any behavioral taxonomies. Even here, however, there are problems as there are no generally accepted taxonomies of learning procedures. How does a researcher interested in studying the intelligence of plants compare and contrast their learning protocols? The answer is they cannot.

Bitterman (1962) and Tulving (1985) persuasively argue that taxonomies can help researchers better characterize and design their learning experiments. Unfortunately, there is no generally accepted taxonomy. Here is another area, in addition to definitional issues related to the term behavior, where the study of plant intelligence can help sharpen how we characterize learning protocols. Taxonomies are available for both classical conditioning (Bitterman 1962; Dyal and Corning 1973; Gormezano and Kehoe 1975) and instrumental/operant conditioning (Bitterman 1962; Woods 1974). The taxonomy proposed by Woods (1974) includes 16 categories of conditioning based on the use of a discriminative stimulus and the attractiveness of the reward.

7 Are Enrichment Theories Needed to Study Plant Learning Ecologically?

In the case of ecological psychology, the question is not whether there are inconsistencies in definitions of learning phenomena or not, but rather whether there is an ecologically valid account of learning and the form it can take. Consider first the cognitivist concept of "enrichment" in mainstream theories of perception. Enrichment theories presuppose poverty-of-the-stimulus arguments (e.g., Chomsky 1980). The relation between the pattern in the energy array at the sensory periphery of an organism and the world is inherently ambiguous, and this is the reason why mainstream cognitivist theories call for an inferential treatment. Perception, for example, is considered to be indirect or mediated, the outcome of a logic-like process of inference required for the sake of reconstructing the world. The classical example is 3D mental reconstruction of 2D retinal images (Marr 1982). Thankfully for plant intelligence research, ecological psychology teaches us that the information available in the environmental is rich enough. Perception has to do with the detection of rich environmental information, not with the enriching of inherently poor stimuli. Ecological psychology, therefore, does not require inner-body entities to stand for environmental objects and properties. If Skinner laid the stress on "contingencies" (unmediated, functional relations) (Morris 2009), ecological perception is equally said to be cognitively unmediated or "direct" (Michaels and Carello 1981). Direct perception gets organized around action, and organisms, plants included, perceive opportunities for action as they interact with their environment in the form of affordances. Some properties of the world match unambiguously the patterns of ambient energy arrays available to a perceptual system in the form of invariant properties of objects, features that remain the same when transformations are applied (Gibson 1966). Organisms pick them up as they move around and explore their environment. According to Gibson, organisms directly resonate with informational invariants that specify opportunities for behavioral interaction in an ever-changing environment. Interestingly, he observes in a revision of the concept of the stimulus in psychology: "For one thing we might search for an invariant component in the bewildering variety of functionally equivalent stimuli. Perhaps there is an invariant stimulus for the invariant response, after all" (Gibson 1960).

One example is the exploration of circumnutation of climbing bean stems (Calvo et al. 2017) under the light of ecological General tau theory (Lee 1998; Lee et al. 2009). A vine does not perceive, say, independent physical variables, such as (absolute) distance, and then infer cognitively the support's availability for twining. In Gibson's view, these absolute variables only make "uninteresting stimuli." But not all physical variables are that *boring*. Consider ratios and proportions. Physics can be ecological, and "if successful, it will provide a basis for a stimulus-response psychology, which otherwise seems to be sinking in a swamp of intervening variables" (Gibson 1960). A vine may instead perceive invariant aspects of the environment that are relevant for biologically relevant interactions. We may say that a plant orienting toward a support behaves in functionally the same way as an animal running toward its prey. Despite things being in constant flux because of the action of twining, some relations remain unchanged, and the plant may pick them up (Calvo et al. 2017). The working hypothesis is that climbing plants resonate to specificational information of the type provided by high-level, relational invariants (Lee et al. 2009), information that can be detected directly, without cognitive mediation.

Many Gibsonians focus on direct perception and turn a blind eye to ecological forms of learning. But direct perception adds to the natural science of perception provided by behaviorism with a natural history of learning that focuses on the invariant context of what is perceived (Morris 2009). Invariant relations can be learned. According to "direct learning" (Jacobs and Michaels 2007), organisms can change an informational variable for another. The question is whether the shift itself is guided directly. When it is, learning, like perception, remains unmediated. We do not need to follow Pavlov and assume an "equivalence of associability assumption" (Seligman 1970; in Michaels and Carello 1981) according to which any association whatsoever may in principle be learned. The ecological scale of interaction between organism and environment constrains what can be learned (something that behaviorism has learned in parallel from the taste aversion literature). Different associations afford different actions. As Michaels and Carello (1981) observe in relation to unnatural settings, "the difficulty observed ... is not that the dogs cannot learn the appropriate behaviors, but that the 'correct' behaviors are not afforded by information." In our view, it is critical that researchers bear this in mind when submitting plants to tests of learning by association. We may approach the learning process as the education of attention (Gibson 1966). Plants are not like animals in this respect.

They resonate to meaningful information, and learn to detect affordances, through structural changes in their nonneural tissues.

8 Is There a Need to Interpret Plant Intelligence in Cognitivist Terms?

We believe that the trend to interpret plant intelligence in cognitivist terms is misplaced. In reading the plant intelligent literature, it seems that little consideration is given to views that compete with the cognitivist position. We believe that this is due to a lack of understanding and interest in knowing about the behaviorist and ecological psychologist positions. As Coffin (1930) wrote almost 90 years ago when discussing the behaviorism of John B. Watson, "So Behaviorism appears as a pathetic figure circling around in the backwash of the widening swiftly flowing stream of science." Sadly this sentiment is still with us being perpetuated by those who have little knowledge of behaviorism (and *mutatis mutandis*, of ecological psychology). Cognitivists perpetuate the myth that cognitive psychology has replaced behaviorism and ecological psychology and this myth has infiltrated the plant intelligence community. But research in plant intelligence can benefit by testing behaviorist and ecological principles in the form of empirical hypotheses subject to experimental scrutiny. We would strongly encourage those interested in the intelligence of plants to consider the behaviorist and the ecological psychology perspectives. Typically, the only behaviorist perspective that cognitivists and those in the plant intelligence community are exposed to are the views of perhaps Watson and B. F. Skinner. The reader may be surprised to learn that there are many forms of behaviorism. In addition to Watson's version and Skinner's "radical behaviorism" are the group of behaviorists known as "neobehaviorists." Neobehaviorists such as Clark Hull, Edward C. Tolman, Abram Amsel, Neal E. Miller, O. H. Mower, and Kenneth W. Spence make use of intervening variables and represent some of the most significant figures in the history of psychology. Pretty much the same goes with respect to ecological psychologists. NeoGibsonians abound, and their views can be as disparate as those of Allan Costall (2011/2012), David N. Lee et al. (2009), or Anthony Chemero (2003, 2009), among others, researchers who have propelled the field in disparate, more or less radical, ways.

Apparently unrecognized by some cognitive psychologists, the use of intervening variables by the neobehaviorists and neoGibsonians shares many characteristics with the cognitive position. Denny (1986), for example, has shown that by altering the meaning of stimulus and response, the cognitive and neobehaviorist position can be merged. Miller (1959) has shown that by modifying some neobehaviorist concepts, motivation and conflict can be better understood. One of the best efforts to reconcile the neobehaviorism of Hull with the cognitive behaviorism of Tolman was undertaken by MacCorquodale and Meehl (1953) who, using a mathematical model, united the views of Hull and Tolman. Stepanov and Abramson (2008) provide a review of early mathematical models associated with neobehaviorism. Leahey (1992) has suggested that as there is little difference between the neobehaviorist position and cognitive psychology, cognitive psychology is not so much so a revolutionary position rather than a successor to neobehaviorism. The reader may likewise be surprised to learn that David Lee has put forward an ecological theory of control, General rho/tau theory (Lee 1998; Lee et al. 2009) that aims to explain how organisms guide their movements endogenously by using prescriptive information.

We are not advocating that those interested in plant intelligence necessarily follow the neobehaviorist and neoGibsonian traditions as some of these traditions can end up being similar to cognitive approaches. What we advocate is the use of the radical behaviorist and ecological positions as represented by Skinner and Gibson, respectively. In both cases, the focus is on the functional analysis of behavior where a researcher looks for observable relationships (Moore 1996, 2011, 2013a, b). Rather than become bogged down in a sea of terms which have no consistent meaning, a functional analysis searches for relationships between independent and dependent variables (Lee 1988). When a relevant independent variable is initially discovered, the effect is systematically replicated—replication is the foundation of a functional analysis (Sidman 1960). Unfortunately, replication is apparently becoming rare in the behavioral sciences with devastating results not only in regard to the lack of replication but also to the way data is analyzed (Grice 2011; Grice et al. 2012).

9 Conclusions and Recommendations

We believe we have identified problems that must be faced sooner or later (preferably sooner) by those interested in exploring the possibility of intelligence in plants. How can one experimentally study, draw conclusions, and make comparisons with other life forms about the intelligence of plants if there are no consistent definitions of "intelligence," "behavior," and "cognition"? The problem is further complicated in that there are no consistent definitions of learning paradigms. After all, it is these paradigms that provide the data demonstrating plant intelligence. One only needs to look at the literature on tool use in animals to see the problems of interpretation that follows when there are no consistent definitions (Crain et al. 2013).

Some researchers may argue that understanding the intelligence of plants can proceed without having consistent definitions of the phenomena they study. We disagree. Moreover, these definitions must be based on data rather than constructing them a priori and then looking for examples that represent the definition (Schlinger 2003). The danger of basing the study of plant intelligence on definitions that are not first based on data is that they soon take on a life of their own and researchers are fooled into believing that their definitions reflect something real. Terms such as intelligence and cognition, for example, are catch phrases that loosely represent various behaviors emitted in a specific context. Unless we can do better in the future, why use such terms? It would seem more parsimonious to eliminate such terms and describe the procedures used to generate the behavior of interest.

In addition to definitional issues, we would advocate that those interested in the intelligence of plants take a look at what the behaviorist and ecological psychology approaches have to offer. Many cognitive psychologists like to believe that their view of intelligence and behavior replaced what they consider to be antique (Moore 1996). Nothing can be further from the truth (Abramson 2013).

We suggest that plant researchers take a look at the radical behaviorism of B. F. Skinner and the ecological psychology of J. J. Gibson where the focus is on discovering functional relationships and not on some hypothetical intervening variable or mentalistic concepts. Parametric studies need to be performed where such parameters as stimulus intensity, CS-US intervals, amount of reward, etc. need to be investigated. Goddard (2012) has suggested that radical behaviorism is making a comeback in psychology. NeoGibsonian psychology is likewise on the rise.

In closing, we support the search for intelligence in plants. Yet our support is tempered by the realization that the plant intelligence community is being led uncritically into problems because of an emphasis on the study of cognition in plants without looking at alternative viewpoints. We hope that this chapter will stimulate at least some of those interested in plant intelligence to better understand the dangers.

References

Abramson CI (1994) A primer of invertebrate learning: the behavioral perspective. American Psychological Association, Washington, DC

Abramson CI (1997) Where have I heard it all before: some neglected issues of invertebrate learning. In: Greenberg G, Tobach E (eds) Comparative psychology of invertebrates: the field and laboratory study of insect behavior. Garland Publishing, New York, pp 55–78

Abramson CI (2013) Problems of teaching the behaviorist perspective in the cognitive revolution. Behav Sci 3:55–71

Abramson CI, Chicas-Mosier AM (2016) Learning in plants: lessons from *Mimosa pudica*. Front Psychol 7:417

Abramson CI, Lack CW (eds) (2014) Psychology gone astray: a selection of the racist and sexist literature from early psychological research. Onus Books, Fareham

Abramson CI, Place AJ (2005) A note regarding the word "Behavior" in glossaries of introductory textbooks and encyclopedia. Percept Motor Skills 101:568–574

Adams F (2010) Why we still need a mark of the cognitive. Cogn Syst Res 11:324–331

Adams F, Aizawa K (2008) The bounds of cognition. Blackwell, Oxford

Affifi R (2013) Learning in plants: semiosis between the parts and the whole. Biosemiotics 6:547–559

Aizawa K (2014) Tough times to be talking systematicity. In: Calvo P, Symons J (eds) The architecture of cognition: rethinking Fodor and Pylyshyn's systematicity challenge. MIT Press, Cambridge, MA, pp 77–99

Alpi A, Amrhein N, Bertl A, Blatt MR, Blumwald E, Cervone F, Dainty J, De Michelis MI, Epstein E, Galston AW, Goldsmith MH, Hawes C, Hell R, Hetherington A, Höfte H, Juergens G, Leaver CJ, Moroni A, Murphy A, Oparka K, Perata P, Quader H, Rausch T, Ritzenthaler C, Rivetta A, Robinson DG, Sanders D, Scheres B, Schumacher K, Sentenac H, Slayman CL, Soave C, Somerville C, Taiz L, Thiel G, Wagner R (2007) Plant neurobiology: no brain, no gain? Trends Plant Sci 12:135–136

Amsel A (1989) Behaviorism, neobehaviorism and cognitivism in learning theory: historical and contemporary perspectives. LEA, Hillsdale, NJ

Baluška F, Levin M (2016) On having no head: cognition throughout biological systems. Front Psychol 7:902

Baluška F, Mancuso S (2009a) Deep evolutionary origins of neurobiology: turning the essence of 'neural' upside-down. Comm Integr Biol 2:60–65

Baluška F, Mancuso S (2009b) Plants and animals: convergent evolution in action? In: Baluška F (ed) Plant-environment interactions: from sensory plant biology to active plant behavior. Springer, Berlin, pp 285–301

Barlow PW (2006) Charles Darwin and the plant root apex: closing a gap in living systems theory as applied to plants. In: Baluška F, Mancuso S, Volkmann D (eds) Communication in plants: neuronal aspects of plant life. Springer, Berlin, pp 37–51

Bitterman ME (1962) Techniques for the study of learning in animals: analysis and classification. Psychol Bull 59:81–93

Brenner ED, Stahlberg R, Mancuso S, Baluška F, Van Volkenburgh E (2007) Response to Alpi et al. plant neurobiology: the gain is more than the name. Trends Plant Sci 12:285–286

Calvo P (2016) The philosophy of plant neurobiology: a manifesto. Synthese 193:1323–1343

Calvo Garzón P, Keijzer F (2011) Plants: adaptive behavior, root-brains, and minimal cognition. Adapt Behav 11:155–171

Calvo P, Martín E, Symons J (2014) The emergence of systematicity in minimally cognitive agents. In: Calvo P, Symons J (eds) The architecture of cognition: rethinking Fodor and Pylyshyn's systematicity challenge. MIT Press, Cambridge, MA, pp 97–434

Calvo P, Baluška F, Sims A (2016) "Feature detection" vs. "predictive coding" models of plant behavior. Front Psychol 7:1505

Calvo P, Raja V, Lee DN (2017) Guidance of circumnutation of climbing bean stems: an ecological exploration. bioRxiv. https://doi.org/10.1101/122358

Carello C, Vaz D, Blau JJC, Petrusz S (2012) Unnerving intelligence. Ecol Psychol 24:241–264

Chaney DW (2013) An overview of the first use of the term cognition and behavior. Behav Sci 3:143–153

Chemero A (2003) An outline of a theory of affordances. Ecol Psychol 15:181–195

Chemero A (2009) Radical embodied cognitive science. MIT Press, Cambridge, MA

Chomsky N (1980) Rules and representations. Basil Blackwell, Oxford

Coffin JH (1930) Can a behaviorist be good? In: King WP (ed) Behaviorism: a battle line! Cokesbury Press, Nashville, TN, pp 242–256

Costall A (1984) Are theories of perception necessary? A review of Gibson's the ecological approach to visual perception. J Exp Anal Behav 41(1):109–115

Costall A (2004) From Darwin to Watson (and cogniti- vism) and back again: the principle of animal-environment mutuality. Behav Philos 32:179–195

Costall A (2011/2012) The hope of a radically embodied science. Behav Philos 39(40):345–353

Crain B, Giray T, Abramson CI (2013) A tool for every job: assessing the need for a universal definition of tool use. Int J Comp Psychol 26:281–303

Cromwell HC, Panksepp J (2011) Rethinking the cognitive revolution from a neural perspective: how overuse/misuse of the term "cognitive" and neglect of affective controls in behavioral neuroscience could be delaying progress in understanding the BrainMind. Neurosci Biobehav Rev 35:2026–2035

Cvrčková F, Lipavská H, Žárský V (2009) Plant intelligence: why, why not or where? Plant Signal Behav 4:394–399

Cvrčková F, Žárský V, Markos A (2016) Plant studies may lead us to rethink the concept of behavior. Front Psychol 7:622

Darwin C (1859) On the origin of species. John Murray, London

Darwin C (1875) The movements and habits of climbing plants. John Murray, London

Darwin C, Darwin F (1880) The power of movement in plants. John Murray, London

Denny MR (1986) "Retention" of S-R in the midst of the cognitive invasion. In: Kendrick DF, Rilling ME, Denny MR (eds) Theories of animal memory. Lawrence Erlbaum Associates, Hillsdale, NJ, pp 5–50

Dyal JA, Corning WC (1973) Invertebrate learning and behavioral taxonomies. In: Corning WC, Dyal JA, Willows AOD (eds) Invertebrate learning: protozoans through annelids, vol 1. Plenum, New York, pp 1–48

Firn R (2004) Plant intelligence: an alternative viewpoint. Ann Bot 93:345–351

Gagliano M, Renton M, Depczynski M, Mancuso S (2014) Experience teaches plants to learn faster and forget slower in environments where it matters. Oecologia 175:63–72

Gagliano M, Vyazovskiy VV, Borbély AA, Grimonprez M, Depczynski M (2016) Learning by association in plants. Sci Rep 6:38427

García Rodríguez A, Calvo Garzón P (2010) Is cognition a matter of representations? Emulation, teleology, and time-keeping in biological systems. Adapt Behav 18:400

Gardner H (2006) Multiple intelligences: new horizons in theory and practice. Basic Books, New York

Gibson JJ (1960) The concept of the stimulus in psychology. Am Psychol 15:694–703

Gibson JJ (1966) The senses considered as perceptual systems. Houghton Mifflin, Boston, MA

Gibson JJ (1979) The ecological approach to visual perception. Houghton Mifflin, Boston, MA

Goddard MJ (2012) On certain similarities between mainstream psychology and the writings of B. F. Skinner. Psychol Rec 62:563–576

Gormezano I, Kehoe EJ (1975) Classical conditioning: some methodological-conceptual issues. In: Estes WK (ed) Handbook of learning and cognitive processes. conditioning and behavior theory, vol 2. Erlbaum, Hillsdale, NJ, pp 143–179

Grice JW (2011) Observation oriented modeling: analysis of cause in the behavioral sciences. Academic, San Diego, CA

Grice J, Barrett P, Schlimgen L, Abramson CI (2012) Toward a brighter future for psychology as an observation oriented science. Behav Sci 2:1–22

Jacobs D, Michaels C (2007) Direct learning. Ecol Psychol 19:321–349

Karban R, Orrock JL, Preisser EL, Sih A (2016) A comparison of plants and animals in their responses to risk of consumption. Curr Opin Plant Biol 32:1–8

Karpinski S, Szechynska-Hebda M (2010) Secret life of plants: from memory to intelligence. Plant Signal Behav 5:1384–1390

Kutschera U, Niklas KJ (2009) Evolutionary plant physiology: Charles Darwin's forgotten synthesis. Naturwissenschaften 96:1339–1354

Leahey TH (1992) The mythical revolutions of American psychology. Am Psychol 47:308–318

Lee VL (1988) Beyond behaviorism. LEA, Hillsdale, NJ

Lee DN (1998) Guiding movement by coupling taus. Ecol Psychol 10:221–250

Lee DN, Bootsma RJ, Frost BJ, Land M, Regan D (2009) General tau theory: evolution to date. Perception 38:837–858

Legg S, Hutter M (2007) A collection of definitions of intelligence. Front Artif Intell Appl 157:17–24

Levitis DA, Lidicker WZ, Freund G (2009) Behavioural biologists don't agree on what constitutes behaviour. Anim Behav 78:103–110

MacCorquodale K, Meehl PE (1953) Preliminary suggestions as to a formalization of expectancy theory. Psychol Rev 60:55–63

Marder M (2013) Plant intentionality and the phenomenological framework of plant intelligence. Plant Signal Behav 8:e23902

Marr D (1982) Vision. MIT Press, Cambridge, MA

Michaels CF, Carello C (1981) Direct perception. Prentice-Hall, New Jersey

Miller NE (1959) Liberalization of basic S-R concepts: extension to conflict behavior, motivation, and social learning. In: Koch S (ed) Psychology: a study of a science, vol 2. McGraw-Hill, New York, pp 196–292

Moore J (1996) On the relation between behaviorism and cognitive psychology. J Mind Behav 17:345–368

Moore J (2011) Behaviorism. Psychol Rec 61:449–464

Moore J (2013a) Tutorial: cognitive psychology as a radical behaviorist views it. Psychol Rec 63:667–680

Moore J (2013b) Three views of behaviorism. Psychol Rec 63:681–692

Morris EK (2009) Behavior analysis and ecological psychology: past, present, and future. A review of Harry Heft's *ecological psychology in context*. J Exp Anal Behav 92:275–304

Open Science Collaboration (2015) Estimating the reproducibility of psychological science. Science 349:4716

Overskeid G (2008) They should have thought about the consequences: the crisis of cognitivism and a second chance for behavior analysis. Psychol Rec 58:131–151

Razran G (1971) Mind in evolution: an east-west synthesis of learned behavior and cognition. Houghton Mifflin, Boston

Retallack DL (1973) The sound of music and plants. DeVorss, Santa Monica, CA

Richardson MJ, Shockley K, Fajen BR, Riley MA, Turvey M (2008) Ecological psychology: six principles for an embodied-embedded approach to behavior. In: Calvo P, Gomila A (eds) Handbook of cognitive science: an embodied approach. Elsevier, Oxford, pp 161–190

Schlinger HD (2003) The myth of intelligence. Psychol Rec 53:15–32

Seligman MEP (1970) On the generality of the laws of learning. Psychol Rev 77:406–418

Sidman M (1960) Tactics of scientific research. Basic Books, New York

Skinner BF (1938) The behavior of organisms: an experimental analysis. Appleton-Century, New York

Skinner BF (1953) Science and human behavior. Macmillan, New York

Skinner BF (1990) Can psychology be a science of mind? Am Psychol 45:1206–1210

Stepanov II, Abramson CI (2008) The application of an exponential mathematical model for 3- arm radial maze learning. J Math Psychol 52:309–319

Sternberg RJ (1984) Toward a triarchic theory of human intelligence. Behav Brain Sci 2:269–315

Sternberg RJ, Detterman DK (eds) (1986) What is intelligence? Contemporary viewpoints on its nature and definition. Ablex, Norwood, NJ

Tolman EC (1932) Purposive behavior in animals and men. Century, New York

Trewavas AJ (2004) Aspects of plant intelligence: an answer to Firn. Ann Bot 93:353–357

Trewavas A (2007) Response to Alpi et al.: plant neurobiology – all metaphors have value. Trends in Plant Sci 12:231–233

Trewavas A (2014) Plant behavior and intelligence. Oxford University Press, New York

Trewavas A (2016) Intelligence, cognition, and language of green plants. Front Psychol 7:588

Tulving E (1985) On the classification problem in learning and memory. In: Nilsson L, Archer T (eds) Perspectives on learning and memory. LEA, Hillsdale, NJ, pp 67–94

Whissell C, Abramson CI, Barber KR (2013) The search for cognitive terminology: an analysis of comparative psychology journals. Behav Sci 3:133–142

Woods PJ (1974) A taxonomy of instrumental conditioning. Am Psychol 29:584–596

Plant Cognition and Behavior: From Environmental Awareness to Synaptic Circuits Navigating Root Apices

František Baluška and Stefano Mancuso

Can Only Animals Know? Why Not Plants?
Karl Popper in All Life is Problem Solving (Popper 1994)
(Page 61 in 2009 Edition)

Abstract Plants emerge as cognitive and intelligent organisms which coevolve with humans since the first flowering plants recognized primates as potential frugivores. Later, when humans started to settle down and initiated the agriculture, our coevolution with crop plants entered a new phase which allowed evolution of our civilization. Here we summarize recent advances in our understanding of plants relying, similarly as animals and humans, on learning and cognition to use their plant-specific behavior for survival. Although plants as such are sessile, their organs move actively and use these movements for active manipulation of their environment, both abiotic and biotic. Moreover, the major strategy of flowering plants is to control their animal pollinators and seed dispersers by providing them with food enriched not only with nutritive but also with manipulative and addictive compounds. There are several examples of cognitive supremacy of plants over animals.

1 Cognition and Behavior in Bacteria, Protozoa, Slime Molds, and Plants

In sciences, optimal strategy is to start your analysis with simple systems and only then to continue to more complex systems. Unfortunately, and in contrast to physics, life sciences dealing with cognition and learning started with more complex systems

---

F. Baluška (✉)
IZMB, University of Bonn, Bonn, Germany
e-mail: baluska@uni-bonn.de

S. Mancuso
LINV, University of Florence, Sesto Fiorentino, Florence, Italy
e-mail: stefano.mancuso@unifi.it

(humans) and only later included also the less complex systems. This unhappy situation is causing fundamental problems and misunderstandings in the current attempts to explore the cognitive behavior of bacteria, protozoa, slime molds, and plants. Fundamental biological phenomena like learning, memory, cognition, sentience, and others were reserved for many years only for humans. Any attempts to expand these basic and fundamental biological concepts to other organisms were, and often still are, dismissed as examples of anthropomorphism (Lyon 2006, 2015).

Nevertheless, cognitive biology is accomplishing a complete Kuhn cycle, when after crisis and revolution phases (Kuhn 1962; Fuller 2003), again normal science phase is emerging in which cognitive sciences already embrace also bacteria, protists, fungi, and plants (Baluška and Mancuso 2009a, b, c; Baluška and Levin 2016; Calvo and Keijzer 2011; Calvo et al. 2017; Gagliano 2017; Gagliano et al. 2017a; Godfrey-Smith 1996; Lyon 2006, 2015; Karpiński and Szechyńska-Hebda 2010; González et al. 2016; Keijzer 2017; Keijzer et al. 2013; van Duijn 2017; Trewavas 2005, 2016, 2017; Trewavas and Baluška 2011; Vallverdú et al. 2018).

Biologically embodied cognition (BEC) is emerging as a new concept unifying organisms at all levels of biological complexity (Lyon 2006, 2015; Baluška and Levin 2016; Keijzer 2017; Vallverdú et al. 2018). Surprisingly, the resistance of mainstream sciences is much stronger in the case of plants (Alpi et al. 2007; Brenner et al. 2007; Trewavas 2007; Biegler 2017; Gagliano et al. 2017b) as in the case of much more simple bacteria or protists. One can only wonder what would be the status of animals should the neurosciences start first with plants, rather than with humans and animals. Could it be that, under such scenario, animals would still not have reached the status of cognitive organisms capable of learning?

2 Land Plants: Behavioral Phase of Flowering Plant Evolution

According to Richard Bateman, there are four major phases of plant evolution: the biochemical phase which ended in the Ordovician (510–438 mya), the anatomical phase spanning Ordovician and Silurian, the morphological phase culminating in the Devonian with fully terrestrialized land plants, and, finally, the behavioral phase of plant evolution that started in the Carboniferous and Permian (359–252 mya) when flowering plants entered into tight coevolution with pollinating insects and other animals, including humans, dispersing plant seeds (Bateman 1991; Bateman et al. 1998). This behavioral phase of flowering land plants apparently speeded up their evolution and evolutionary success, a phenomenon known as Darwin's "abominable mystery" (Crepet and Niklas 2009; Friedman 2009). Sudden origin and fast diversification of flowering land plants still require an explanation. Cognition and behavioral competences of plants were not discussed as relevant factors yet. Interestingly in this respect, orchids represent one of the most species-rich plant families, and they are fooling their pollinators with clever sexual deceits and manipulations (Cozzolino and Widmer 2005; Jersáková et al. 2006; Gaskett et al. 2008; Schiestl 2005, 2010). This supports the view that plant cognition might be linked to plant evolution.

Besides orchids fooling their pollinators using only shapes and colors of their flowers, there are further convincing examples of a cognitive supremacy of flowering plants over their animal-based sexual vectors. For example, there are several examples of behavioral control of pollinators via complex chemistry of their flower nectar. For example, plants use caffeine, alkaloids, and other still not well-understood secondary chemicals which allow plants to manipulate cognition and behavior of their pollinators (Köhler et al. 2012; Irwin et al. 2014; Couvillon et al. 2015; Pyke 2016; Baracchi et al. 2017; Barlow et al. 2017; Koch and Stevenson 2017). Actively changing the chemical composition of nectar allows plants to actively select their pollinators (Kessler and Baldwin 2007; Kessler et al. 2008, 2010). This is reminiscent of acacia trees which attract and manipulate aggressive ants for their protection against herbivores as well as for the seed dispersal (González-Teuber et al. 2014; Mayer et al. 2014). Acacia trees provide ants with nectar which then manipulates their behavior and cognition (Nepi 2014, 2017; Grasso et al. 2015). Diverse neuromodulatory compounds found in nectar not only make ants completely addicted but also can increase their aggressiveness (Szczuka et al. 2013; Grasso et al. 2015), making them even more effective "bodyguards" of plants. In the case of the so-called devil's gardens, *Myrmelachista schumanni* ants are so manipulated that they attack and remove not only all herbivores from their host plants but also all plants around the Amazonian rainforest tree *Duroia hirsuta* (Frederickson et al. 2005; Frederickson and Gordon 2007).

Finally, tomato plants attacked by small mottled willow moth caterpillars release chemicals which turn these caterpillars into cannibals (Orrock et al. 2017). This plant-induced caterpillar cannibalism benefit tomato plants in two different ways. Firstly, cannibalism directly reduces caterpillar abundance. Secondly, cannibalistic caterpillars eat significantly less tomato leafs. Again, the supremacy of plant intelligence over that of the animal herbivores is quite obvious. Plants are using clever strategies to control animals. These strategies are based on fooling and manipulating animals using such diverse set of tools as chemicals (toxins, drugs, odors) but also purely abstract physical cues such as shapes, special surfaces (slippery or sticky insect traps), and colors which not only attract but even immobilize temporarily their prey and pollinators (Oelschlägel et al. 2009; Bonhomme et al. 2011; Bauer et al. 2012; Broderbauer et al. 2013; Henning and Weigend 2013a, b; Phillips et al. 2014). One of the most shocking examples of the plant supremacy over animals was the report about the killing of an African antelope kudu by acacia trees, warned via their over-grazed neighbors releasing ethylene, by increased levels of tannins in their leaves (Van Hoven 1991).

3 Plant Awareness of Their Environment and Neighboring Organisms

Despite such examples, mainstream plant sciences still consider plants for simple and passive organisms placed outside of cognition and behavior realms. Nevertheless, some scientists, even as famous as Charles Darwin, repeatedly reminded us that this view is rather oversimplification of the true nature of plants. Unfortunately, the dominating view in plant sciences is that plants are processing environmental information automatically, without any neuronal-like sensory-motor systems and processes (Alpi et al. 2007; Brenner et al. 2007; Trewavas 2007; Busch and Benfey 2010). However, as it was stated already by Karl Popper, all life is problem solving; and plants, similarly as animals, are well aware of their environment (Popper 1994, pages 61–64 in 2009 Edition). As the awareness of environment is, in essence, consciousness and sentience (Margulis and Sagan 1995; Trewavas and Baluška 2011; Calvo 2017; Calvo et al. 2017), it can be suggested that plant-specific consciousness (Trewavas and Baluška 2011; Gardiner 2012; Calvo et al. 2017) is essential for sensory aspects of plant life as well as for adaptation and survival via plant-specific learning based on memories and predictions (Baluška et al. 2016; Gagliano 2017; Gagliano et al. 2016; Calvo 2017; Calvo et al. 2016; Calvo and Friston 2017). This new view of plants, including the still speculative concept of the plant-specific consciousness, is supported by our recent study showing that diverse anesthetics prevent plant movements via blockage of action potentials (Yokawa et al. 2018). As in animals, also plant movements are obviously animated via plant-specific action potentials (Sibaoka 1991; Grémiaux et al. 2014; Baluška et al. 2016; Böhm et al. 2016; Pavlovič et al. 2017; Yokawa et al. 2018) sensitive to anesthetics (Yokawa et al. 2018). Without having capacities of awareness and behavioral flexibility, plants would not be able to cope with numerous biotic and abiotic stresses to which they are continually exposed due to their sessile nature. One of the first convincing recent evidences of plant senses allowing them to track their neighboring organisms was the discovery that olfactory cues are used by parasitic dodder vine to locate their tomato host (Runyon et al. 2006; Mescher et al. 2006).

Although growing underground, plant roots detect light and use it as a cue for their navigation (Baluška and Mancuso 2016, 2017; Burbach et al. 2012; Mo et al. 2015; Yokawa and Baluška 2015). Too much light is stressful not only for roots but also for the whole plant (Karpinski et al. 1999; Gilroy et al. 2016; Hedrich et al. 2016; Białasek et al. 2017; Szechyńska-Hebda et al. 2010, 2017), and roots do everything possible to minimize amounts of photons hitting their surfaces (Burbach et al. 2012; Yokawa et al. 2011, 2014). Maize roots exposed to light starts to accomplish vigorous root crawling and are more sensitive to gravity, accomplishing U-turns if inserted into thin glass capillaries in inverted position (Burbach et al. 2012; Suzuki et al. 2016). This maize root behavior is based on the light-induced de novo auxin synthesis in root apices (Suzuki et al. 2016). Arabidopsis roots also start to perform root crawling behavior if exposed to light (Mauritz Sommer and Frantisek Baluska, unpublished data). However, even roots growing underground in soil perceive and monitor light from aboveground which controls their growth and root system architecture (Galen et al. 2007a, b; Moni et al. 2015; Lee et al. 2016a, b, 2017).

Arabidopsis and maize roots can also detect and evaluate phonons of sound waves (Gagliano et al. 2012, 2017a; Rodrigo-Moreno et al. 2017). Plant roots are capable to evaluate physical parameters of sound waves and even to locate their source at large distances. Sound waves of 200 Hz, which are in a range of streaming water, induce positive phonotropism of *Arabidopsis* roots via PIN2-based auxin transport and cellular signaling including calcium, potassium, and reactive oxygen species (Rodrigo-Moreno et al. 2017). Intriguingly, growing maize root apices generate periodic clicking sounds (Gagliano et al. 2012) which might be used as some kind of root echolocation that would even inform roots about physical parameters of soil. Moreover, flowers perceive buzz of pollinators (De Luca and Vallejo-Marin 2013; Mishra et al. 2016) and shoots recognize sound waves generated by chewing insect larvae and other herbivores (Gagliano 2013; Appel and Cocroft 2014; Chowdhury et al. 2014; Mescher and De Moraes 2015; Mishra et al. 2016). Obviously, besides olfactory and light cues, also sound cues are emerging to be ecologically and behaviorally relevant for plants.

4 Plant Intelligence: Learning and Cognition Drives Plant Adaptation and Evolution

Plant awareness implicates not only online knowledge of environmental factors and conditions but also some kind of understanding as all sensory stimuli must be integrated and placed into the actual context. The nature of sensory knowledge of plants is critical as the environmental context is also changing continuously. Plant-specific intelligence allows rapid modification of plant behavior and selecting, or prioritizing (Calvo et al. 2017), the most critical aspects of the environment relevant for their effective adaptation and survival. In other words, plants enjoy not only their own plant-specific intelligence (Trewavas 2005, 2014, 2016, 2017) but also their plant-specific versions of attention and intentionality (Marder 2012, 2013) to identify and respond properly with the most relevant environmental parameters. Moreover, plants manipulating and controlling their pollinators and protectors (bodyguards) apparently have perfect knowledge (models) of them which allows plants to keep supremacy over these animals.

5 Avoidance and Escape Behavior in Plants

As a part of the current paradigm shift in cognitive sciences, plant behavior is emerging as valid science slowly accepted by the mainstream plant sciences (Karban 2008; Metlen et al. 2009; Trewavas 2014; Calvo et al. 2016; Cvrčková et al. 2016; van Loon 2016). Plant behavior is embedded within still controversial field of the plant neurobiology (Brenner et al. 2006, 2007; Trewavas 2007) which includes also plant communication, plant intelligence, and plant sentience (Calvo 2017; Calvo and Friston 2017; Calvo et al. 2017; Trewavas 2014, 2016, 2017).

One of the new aspects of plant physiology is active response to adverse situations of mild stress when faster growth of plant organs (especially roots) does not mean, as classical plant physiology posit, optimal growth conditions. For instance, plant root growth is faster after illumination not because roots require light for their maximal growth rate (as the traditional plant physiology would suggest) but rather because roots perceive light as a stress factor and their faster growth is allowing them to escape from this stressful situation (discussed below). As Arabidopsis is cultivated in laboratories within the transparent Petri dishes since the 1970s of the last century, and plants are known to memorize their stress vie epigenetic memory, we can predict that the laboratory-grown Arabidopsis lines are much affected after so many generations having their roots exposed to light (Yokawa et al. 2011, 2014). Even more significantly, as root stress is rapidly spreading within plant tissues and organs, seedlings with light-exposed roots are stressed, and many data obtained with such stressed seedlings will require more or less significant reinterpretations of large amounts of published data.

Avoidance and escape tropisms are known also from shoots which actively avoid dark spots and compete for light (Pierik et al. 2011; Pierik and de Wit 2014; Gundel et al. 2014; Pierik and Testerink 2014; de Wit et al. 2016; Ballaré and Pierik 2017; Pantazopoulou et al. 2017). Until now, shade avoidance is studied mostly from the perspective of plant hormones, especially auxin. But plants perceive and respond to shade also via their action potentials (Stahlberg et al. 2006). Importantly in this respect, glutamate-induced action potentials control shoot growth and circumnutations in Helianthus seedlings (Stolarz and Dziubinska 2017).

One of the most intriguing and relatively well characterized escape tropisms in plants is the salt stress escape tropism of Arabidopsis roots (Li and Zhang 2008; Sun et al. 2008). Growing Arabidopsis roots somehow recognize the salt stress medium of the Petri dish agar plates and avoid this salty patch by degradation of their amyloplast-based statoliths in root cap statocytes and by modifications of the PIN2-driven basipetal (shootward) auxin transport to accomplish root halotropism (Galvan-Ampudia and Testerink 2011; Galvan-Ampudia et al. 2013; Yokawa et al. 2014). This salt area avoidance tropism (halotropism) was confirmed also for tomato and sorghum roots grown in soil (Galvan-Ampudia et al. 2013; Pierik and Testerink 2014). Similar differential degradation of PIN2 also allows escape and avoidance of Arabidopsis roots from low oxygen (hypoxia) patches (Eysholdt-Derzsó and Sauter 2017) and from the high ATP level containing agar medium (Zhu et al. 2017a). This eATP avoidance tropism of Arabidopsis roots was relying on calcium influx into root cells regulated by the activity of heterotrimeric G proteins.

It will be important to test future roles of eATP receptor DORN1 (Choi et al. 2014a) and heterotrimeric G-protein-based calcium influx also in hypoxia and salt stress avoidance tropisms. Importantly in this respect, eATP serves as universal danger signal for plants (Cao et al. 2014; Choi et al. 2014b; Tanaka et al. 2014). Interestingly in this respect, eATP acts in jasmonic acid (JA) signaling (Tripathi et al. 2017) and JA controls endocytosis and vesicle recycling of PIN2 (Sun et al. 2011). JA controls root growth via polar auxin transport in cells of the root apex transition zone (Yan et al. 2016). Intriguingly, even insect herbivory of leaves inhibits root growth by affecting the auxin fluxes at the root apices in *Arabidopsis thaliana* (Yan et al. 2017).

6 Surprises from the Plant Neurobiology Corner

6.1 Root Growth Speeds Up Under Mild Stress and Slows Down After Finding Nutrition

Most of our knowledge on plants and their roots come from the model plant *Arabidopsis thaliana* grown within transparent Petri dishes placed with controlled growth chambers of our laboratories (Koncz 2006; Buell and Last 2010). Classical interpretation of molecular biologists and plant physiologists is that fast growth of plant organs means optimal growth conditions whereas slow growth results from stress conditions. These traditional views, based fundamental differences between plants and animals, are derived from the still dominating view that plants are some kind of automata, or physiological zombies, being deprived of any mental and cognitive faculties. The alternative scenario provided by plant neurobiology is that plants and animals are based on the same neurobiological principles (Brenner et al. 2006, 2007; Baluška and Mancuso 2007, 2009a, b, c; Baluška 2010). Arabidopsis seedlings are cultivated in the transparent Petri dishes on a routine basis in all laboratories throughout the globe since 1970s. This huge amount of data obtained with such experimental system will require some reinterpretation in the future.

6.2 Root Foraging Behavior

The lifestyle of plant roots is very active as they act in a very complex and dangerous environment of soil. In order to provide plants with water and nutrition, roots team up with mycorrhiza fungi and numerous plant growth-promoting bacteria. As water and nutrients are distributed in patchy manner, roots actively search for nutrients and avoid dry and toxic soil areas. After reaching nutritionally rich areas, roots stop their growth and produce numerous lateral roots in order to exploit such areas (Giehl and von Wirén 2014). While searching and escaping root behavior is associated with vigilant root growth, foraging and patch exploiting behavior is based on roots which do not grow fast but rather generate new roots. After the patch is depleted in nutrients, roots regain the searching behavior based on rapid root growth. Root hair tip growth is also associated with nutrient acquisition and less so with rapid root growth during the searching or escaping phases of root behavior. In the case of emergency, it is better to invest energy into the root growth to find rapid water/nutrients or to escape dangerous stress situations (Baluška et al. 2009b; Barrada et al. 2015; Yokawa and Baluška 2016).

However, the life of roots is determined via more complex situations when, for example, nutritionally rich areas are often connected with other abiotic and/or biotic challenges. Such more demanding situations require more complex root cognition in order to choose the most optimal behavior strategy (Kawa et al. 2016). Root apices growing in soil and continuously "bombarded" with information from their sensory systems need to integrate this complex to make proper decisions (Hodge 2009). PIN2 is critical not only for escape and avoidance tropisms but also for proper root decisions with respect to searching for phosphate and nitrate nutrition (Liu et al. 2013; Zou et al. 2013; Pandya-Kumar et al. 2014; Kumar et al. 2015; Koltai 2015).

6.3 Roots Actively Manipulate Soil via Chemical Exudations: Creating Survival Niche

Roots actively recruit friendly bacteria and fungi. For example, Arabidopsis roots selectively secrete malic acid to recruit beneficial rhizobacteria (Rudrappa et al. 2008; Bulgarelli et al. 2012; Kumar and Bais 2012; Foo et al. 2013; Kobayashi et al. 2013a, b; Lakshmanan et al. 2013). Moreover, plant roots also secrete numerous secondary metabolites that attract root growth promoting bacteria (Knee et al. 2001; Walker et al. 2003; Badri and Vivanco 2009; Carvalhais et al. 2015), symbiotic Rhizobium bacteria, and arbuscular mycorrhizal fungi (Kent Peters and Long 1988; Besserer et al. 2006). For example, root exudate strigolactones into their rhizosphere to attract symbiotic arbuscular mycorrhizal fungi (Akiyama and Hayashi 2006; López-Ráez et al. 2017; Waters et al. 2017). In addition, similarly as in the case of shoots, also herbivory-attacked roots release infochemicals attracting enemies (bodyguards) of these herbivores (Rasmann et al. 2005; Degenhardt et al. 2009). Interestingly, strigolactones not only attract symbiotic bacteria and fungi but also connect root colonization by mycorrhizal fungi to phosphate availability (Foo et al. 2013) and mediate plant-to-plant communication between parasitic angiosperms and their hosts (Estabrook and Yoder 1998; Waters et al. 2017).

Besides the attractions of beneficial bacteria and fungi, roots also secrete abundant exudates to actively modify chemical composition of soil to chelate toxic substances and release critical mineral nutrition from more complex chemicals (Raghothama 1999; Tsai and Schmidt 2017). Finally, root exudates actively control and shape physical and chemical properties of soil (Morel et al. 1991; Watt et al. 1994; Hoffland et al. 1989; Lapsansky et al. 2016; Galloway et al. 2017).

Finally, roots use their exudates to recognize roots from other plants, either from the same plant, roots of another plant of the same species, or roots of a plant from another species (Gruntman and Novoplansky 2004; Falik et al. 2006; Novoplansky 2009; Biedrzycki and Bais 2010; Biedrzycki et al. 2010; Bhatt et al. 2011; Chen et al. 2012). Roots will use this information to decide about their future behavior (Murphy and Dudley 2009; Novoplansky 2009; Chen et al. 2012).

6.4 Roots Stop Active Growth After Experiencing Low Phosphate and Aluminum Toxicity

Recently, important breakthrough was achieved in our understanding of root growth inhibition due to the low phosphate availability. It turned out to be an active root growth stop (some kind of a root growth brake) mediated by the STOP1-ALMT1 signaling (Abel 2017; Balzergue et al. 2017; Mora-Macías et al. 2017). Intriguingly, the STOP1-ALMT1 signaling is used also to stop root growth actively if root apices experience aluminum toxicity (Kobayashi et al. 2013a, b; Fan et al. 2016; Daspute et al. 2017). If the STOP1-ALMT1 signaling is compromised, then such roots are not able to stop their growth even if they are exposed to the low phosphate or high aluminum environment (Abel 2017; Balzergue et al. 2017; Mora-Macías et al. 2017; Fan et al. 2016; Daspute et al. 2017). In other words, roots do not stop to grow under low phosphate or high aluminum due to some direct impacts of their toxicity but rather as a result of active "decision" to stop the root growth as it can be dangerous for such root to continue its growth. Interestingly in this respect, both low phosphate and high aluminum are sensed directly at the root apex (Sivaguru et al. 1999, 2003; Svistoonoff et al. 2007). Recent studies suggest that the jasmonic acid signaling will be involved in this active ("deliberate") stop of root growth under salt stress and aluminum toxicity (Valenzuela et al. 2016; Yang et al. 2017). Interestingly in this respect, leaf herbivory also inhibits root growth via jasmonic acid—mediated targeting of auxin fluxes in cells of the root apex transition zone (Yan et al. 2017).

7 Root Cap as Sense Organ for Root Navigation: Root Cap—PIN2 Connection

Since the sedimenting statoliths of the root cap statocytes were discovered to act as plant gravity sensors for root gravitropism (Haberlandt 1900; Nemec 1900, 1902), mysterious transmission of gravity signals from the root caps sensing gravity to the elongation zone, accomplishing root tropisms, was discussed extensively (Sievers et al. 1984; Barlow 1995; Konings 1995; Baluška and Hasenstein 1997; Monshausen et al. 1996; Plieth and Trewavas 2002; Fasano et al. 2002; Wolverton et al. 2002; Perrin et al. 2005; Hahn et al. 2006; Baluška et al. 2007). Root cap was known to play a crucial role in this long-distance transmission of gravity signals since Charles and Francis Darwin concluded from their surgical experiments that maize roots devoid of root caps continue to grow but fail to accomplish root tropisms (Darwin 1880; Baluška and Mancuso 2009c; Baluška et al. 2009a). Interestingly in this respect, maize roots devoid of their root caps are not able to sense ethylene (Hahn et al. 2008). Moreover, the PIN2 mutant of Arabidopsis, which is also devoid of any root tropisms, was identified originally as ethylene-insensitive root mutant (Chen et al. 1998; Dolan 1998; Müller et al. 1998; Luschnig et al. 1998; Utsuno et al. 1998). Similarly as decapped maize roots, Arabidopsis roots of the PIN2 mutant grow vigorously but are unable to navigate their growth direction with respect to any of the otherwise relevant environmental cues. In other words, such roots still able to grow but they are *blind* to their surroundings. Obviously, the PIN2 protein plays a central role in the sensory root cap-based root behavior. PIN2 protein is having a crucial role in the Darwinian root apex brain. Another very interesting PIN with respect to the Darwinian root apex brain theory is PIN6 which is expressed highly in nectaries and root apices, exerts inhibitory role on the root growth, and controls ethylene sensitivities (Bender et al. 2013; Cazzonelli et al. 2013; Nisar et al. 2014; Ditengou et al. 2017).

Moreover, PIN6 is important also for cellular auxin homeostasis (Cazzonelli et al. 2013; Simon et al. 2016) and relevant for the flower development (Ditengou et al. 2017). Another relevant aspect in this respect is that PIN6 is controlled via MPK6 phosphorylation (Ditengou et al. 2017). MPK6 is expressed strongly in the PIN2 expression domain of the root apex transition zone (Müller et al. 2010), and MPK6 also controls auxin levels in root apices (Smékalová et al. 2014) and cytoskeleton as well as PLD (Takáč et al. 2016), known to be driving endocytic vesicle recycling relevant for the auxin fluxes in the root apex transition zone (Mancuso et al. 2007). Similarly as PIN6, also MPK6 acts as repressor of the root growth (López-Bucio et al. 2014; Smékalová et al. 2014; Contreras-Cornejo et al. 2015; Takáč et al. 2016).

Besides the shootward (basipetal) polar auxin flow, also PIN2-mediated control of the plasma membrane H^+-ATPase activity emerges as relevant process under the control of PIN2 (Xu et al. 2012). PIN2 is showing unique properties among all other PINs via its unique conserved cysteines and regulation via ROS and NO reactive species (Ni et al. 2017; Retzer et al. 2017; Qu et al. 2017). Importantly, NO signaling is very active and connected to auxin transport in cells of the root apex transition zone in which PIN2 expression is at maximum (Schlicht et al. 2008; Baluška et al. 2010; Baluška and Mancuso 2013; Manoli et al. 2014, 2016; Trevisan et al. 2014, 2015). Another unique aspect of PIN2, making it rather unique from all other PINs expressed in root apices (PIN1, PIN3, PIN4, PIN5, PIN6, PIN7), is its very high biosynthesis and accumulation of newly synthesized PIN2 within the endocytic BFA-induced compartments (Jásik et al. 2016).

8 Plant Sensory Systems Feed into Dynamic Plant Synapses to Guide Root Behavior

How can we explain plant cognition and plant behavior at the cellular level? In the last two decades, the popular belief that plants lack neurons and synapses is emerging to be wrong, and plant cells have almost all the features which were considered to be specific only to neurons (Baluška 2010) of animal/human brains. One can envision that synaptic cell-cell adhesions at the crosswalls are processing sensory information in a process which feeds back to endocytic vesicle recycling modulating the strengths of these synaptic cell–cell adhesions (Baluška and

Mancuso 2013; Baluška et al. 2005a, 2009a, b). This process is remotely similar to synaptic plasticity of neuronal synapses in brains of animals, underlying memory formation, learning, and behavior of animals (Tononi and Cirelli 2014; Nonaka et al. 2014). For example, associative learning in brain is based on glutamate receptor trafficking (Rumpel et al. 2005). Dynamic actin cytoskeleton is crucial for the synaptic plasticity related to sensory information processing and storage in the brain (Hotulainen and Hoogenraad 2010; Bellot et al. 2014; Lei et al. 2017; Yasuda 2017). Similar role for the actin cytoskeleton can also be expected for control of synaptic plasticity in plants. In the future, it is necessary to combine studies on the plant cytoskeleton with associative learning (Gagliano et al. 2016). For example, the polar cell-cell transport of auxin is known to be under the control of the actin cytoskeleton (Dhonukshe et al. 2008; Nick 2010; Nagawa et al. 2012; Zhu and Geisler 2015; Zhu et al. 2017b; Eggenberger et al. 2017; Huang et al. 2017), whereas root behavior is based on the polar auxin transport (Baluška et al. 2005a, 2009a, b; Baluška and Mancuso 2013). Surprisingly, insect leaf herbivory inhibits root growth via JA—mediated targeting of auxin flux in cells of the root apex transition zone (Yan et al. 2017). Besides auxin fluxes, leaf herbivory affected also H^+ fluxes via stimulating of the plasma membrane H^+-ATPase activity. Moreover, expression of PIN1, PIN2, PIN3, PIN7, and AUX1 auxin transporters were downregulated in root apices, whereas transcripts for auxin biosynthesis YUCCA genes were upregulated (Yan et al. 2017). In other words, insect attack and herbivory of shoot leaves elicits responses in root synapses of the root apex transition zone (Baluška et al. 2005a, 2009a, b; Baluška and Mancuso 2013). Similarly, shoot herbivory was also reported to induce, via airborne volatiles as well as via ALMT1-mediated secretion of malic acid from root apices, auxin responses in the root apices of neighboring plants (Sweeney et al. 2017).

9 Auxin, Glutamate, and GABA: Emerging Transmitters in Plants

Polar cell-cell transport of auxin in root apices, with maxima at the transition zone, are based on synaptic activities (Mancuso et al. 2005, 2007; Baluška et al. 2005a, 2009a, b; Baluška and Mancuso 2013). Importantly in this respect, classical auxin transport inhibitors inhibit endocytic vesicle recycling (synaptic activity, Mancuso et al. 2007; Baluška et al. 2005a, 2008), and auxin seems to be localized within recycling vesicles in cells of transition zone (Schlicht et al. 2006; Mettbach et al. 2017). Moreover, in support of the vesicular auxin in plants, direct detection of quantal IAA release from plant cell protoplasts in real time was reported using amperometry based on microelectrodes decorated with the IAA-specific nanowires (Liu et al. 2014).

In plant root apices, both glutamate and GABA are emerging to act as similar transmitters as in brains of animals and humans. Arabidopsis, and obviously also other plants, use glutamate and GABA to control excitability of their plasma membranes via glutamate-like receptors (Dubos et al. 2003; Michard et al. 2011; Price et al. 2012; Weiland et al. 2016; Ortiz-Ramírez et al. 2017) and GABA receptors (Ramesh et al. 2015, 2017; Žárský 2015). Animal and human learning via the synaptic plasticity is based on the endocytic recycling of glutamate and GABA receptors (Gu and Huganir 2016; Gu et al. 2016; Chiu et al. 2017). Molecular basis of synaptic recycling at animal/human and plant synapses is almost identical (Šamaj et al. 2005; Mancuso et al. 2007; Luschnig and Vert 2014; Sekereš et al. 2015; Tu-Sekine et al. 2015; Barber et al. 2017). GABA and glutamate activate their plant plasma membrane receptors from the external cell wall space, suggesting that they will be also released from neighboring cells via regulated vesicular secretion. Thus, besides auxin, glutamate and GABA also emerge to act as transmitters at plant synapses, with potential roles in guiding the exploratory root growth in soil.

10 Transition Zone with Active Plant Synapses as Plant Brain-Like Command Center

PIN2 protein is expressed preferentially in cells of the root apex transition zone and undergoes continuous vesicular cycling under the transversal cell walls (or end poles) of epidermis and cortex cells (Schlicht et al. 2008; Wan et al. 2012; Jásik et al. 2013). These actin cytoskeleton-based cell-cell adhesion domains fulfill all the criteria of synaptic domains showing several strong analogies, including endocytic vesicle recycling controlled by actin cytoskeleton and underlying cell-cell transport of transmitters, with the neuronal and immunological synapses of animals or humans (Baluška et al. 2003, 2005a, 2009a, b). These plant synapses are very dynamic as also cell wall materials, particularly pectins, cross-linked with calcium and boron as well as xyloglucans, accomplishing endocytic vesicle recycling (Baluška et al. 2002, 2005b; Dhonukshe et al. 2006; Manoli et al. 2016). These dynamic cell-cell adhesion domains, based on the actin cytoskeleton-driven cell wall pectins recycling, are very sensitive to environmental and endogenous cues via their plant-specific endocytic vesicle recycling apparatus (Baluška et al. 2003, 2005b) which includes also plant-specific myosin VIII and formins involved in plant endocytosis (Volkmann et al. 2003; Baluška and Hlavacka 2005; Golomb et al. 2008; Sattarzadeh et al. 2008). This allows endocytic recycling of membranes and critical molecules, as well as cell wall components, making these plant synapses excellent cognitive structures guiding root growth via processing of sensory information at the root apex synapses (Baluška et al. 2009a,b). All these data fit nicely into the concept of root brain originally proposed by Charles and Francis Darwin already in 1880 (Darwin 1880; Barlow 2006; Baluška et al. 2004a, 2009b; Kutschera and Niklas 2009). Recent study revealed that general and local anesthetics block plant movements via targeting of the synaptic endocytic vesicle recycling and blocking action potentials (Yokawa et al. 2018).

11 Transition Zone as Command Center for Root Navigation: Energides in Driver Seats?

It is not often appreciated but root apices, pushed forward by their elongation region, cover very large distances. Dittmer (1937) analyzed and calculated the root system of a single rye plant grown in the field. The total lengths of all roots of this single rye plant covered 622 km which means that individual root apices traverse more than several meters in one season. This is a tremendous feat if you realize that the rye root apex diameter is less than 1 mm. What kind of navigation system is achieving such a performance? The root apex transition zone is in a perfect position and having optimal properties in this respect (Baluška et al. 2009b, 2010; Baluška and Mancuso 2013).

Plant synapses are communicating with nuclei as evidenced by rather dramatic reposition of transition zone nuclei after brefeldin A-induced inhibition of synaptic activity (see Fig. 1 in Baluška and Hlavacka 2005). Similarly, synaptic activity controls geometry of nuclei in neurons of rat hippocampus (Wittmann et al. 2009). Root apex transition zone represents a hotspot of the electric spiking (Masi et al. 2009), and these plant-specific action potentials are in perfect position to control root navigation via plant synapses closely interlinked with nuclei. Moreover PIN2, which is the critical PIN with respect to all root tropisms, is expressed together with PIN6 and MPK6 in cells of the root apex transition zone (discussed above), showing the highest levels of auxin fluxes (Mancuso et al. 2005, 2007). Importantly, MPK6 co-localizes with endosomes, TGNs, endosomal vesicles, and perinuclear microtubules (Müller et al. 2010). The nuclei and their microtubules, also termed as Energides—cell bodies (Baluška et al. 2004b, 2012)—are emerging as descendant of ancient symbiotic events (Baluška and Lyons 2018a, b). Inhibition of endocytic vesicle recycling with brefeldin A inhibits these transition zone peak auxin fluxes (Mancuso et al. 2005, 2007), as well as special arrangements of perinuclear F-actin bundles (Baluška et al. 1997; Baluška and Hlavacka 2005). Similarly as in the case of PIN2 mutant roots of Arabidopsis, or decapped maize roots, also brefeldin A-exposed roots are unable to accomplish root tropisms. Therefore, an exciting scenario is emerging according to which these Energides—plant synapse interactions—are acting together to drive the root apex navigation via sensory perceptions obtained from the root caps, as well as from other root apex cells. The Darwinian root apex brain theory (Darwin 1880; Barlow 2006; Baluška et al. 2009b; Baluška and Mancuso 2013) is shaping up.

12 Conclusions

Plant behaviors based on plant sensing, learning, memorizing, and communicating are inherent features of plant life (Trewavas 2014; Karban 2015). In many aspects, the flowering plants even outsmart animals (Van Hoven 1991; Lev-Yadun 2016a,

2016b, 2017; Orrock et al. 2017) using swindle and manipulation in order to control animals as their survival strategy (Schiestl 2005; Nepi 2014, 2017; Grasso et al. 2015). Similarly as animals, also plants are *animating* movements of their organs via action potentials which are sensitive to anesthetics. In this respect, we should be prepared to more surprises in the future. As our life on this planet Earth, and in fact the whole modern civilization, depends completely on plants, we should start to study plants also from their neuronal, sensory, and cognitive aspects.

References

Abel S (2017) Phosphate scouting by root tips. Curr Opin Plant Biol 39:168–177

Akiyama K, Hayashi H (2006) Strigolactones: chemical signals for fungal symbionts and parasitic weeds in plant roots. Ann Bot 97:925–931

Alpi A, Amrhein N, Bertl A, Blatt MR, Blumwald E, Cervone F, Dainty J, De Michelis MI, Epstein E, Galston AW, Goldsmith MH, Hawes C, Hell R, Hetherington A, Hofte H, Juergens G, Leaver CJ, Moroni A, Murphy A, Oparka K, Perata P, Quader H, Rausch T, Ritzenthaler C, Rivetta A, Robinson DG, Sanders D, Scheres B, Schumacher K, Sentenac H, Slayman CL, Soave C, Somerville C, Taiz L, Thiel G, Wagner R (2007) Plant neurobiology: no brain, no gain? Trends Plant Sci 12:135–136

Appel HM, Cocroft RB (2014) Plants respond to leaf vibrations caused by insect herbivore chewing. Oecologia 175:1257–1266

Badri DV, Vivanco JM (2009) Regulation and function of root exudates. Plant Cell Environ 32:666–681

Ballaré CL, Pierik R (2017) The shade-avoidance syndrome: multiple signals and ecological consequences. Plant Cell Environ 40:2530–2543

Baluška F (2010) Recent surprising similarities between plant cells and neurons. Plant Signal Behav 5:87–89

Baluška F, Hasenstein KH (1997) Root cytoskeleton: its role in perception of and response to gravity. Planta 203:S69–S78

Baluška F, Hlavacka A (2005) Plant formins come to age: something special about cross-walls. New Phytol 168:499–503

Baluška F, Levin M (2016) On having no head: cognition throughout biological systems. Front Psychol 7:902

Baluška F, Lyons S (2018a) Symbiotic origin of eukaryotic nucleus—from Cell Body to Neo-Energide. In: Sahi VP, Baluška F (eds) Concepts in cell biology: history and evolution. Springer, Berlin. (In press)

Baluška F, Lyons S (2018b) Energide-cell body as smallest unit of eukaryotic life. Ann Bot. (In press)

Baluška F, Mancuso S (2007) Plant neurobiology as a paradigm shift not only in the plant sciences. Plant Signal Behav 2:205–207

Baluška F, Mancuso S (2009a) Plant neurobiology: from sensory biology, via plant communication, to social plant behaviour. Cogn Process 10(Suppl. 1):3–7

Baluška F, Mancuso S (2009b) Deep evolutionary origins of neurobiology: turning the essence of 'neural' upside-down. Commun Integr Biol 2:60–65

Baluška F, Mancuso S (2009c) Plants and animals: convergent evolution in action? In: Baluška F (ed) Plant-environment interactions: from sensory plant biology to active plant behavior. Springer, Berlin, pp 285–301

Baluška F, Mancuso S (2013) Root apex transition zone as oscillatory zone. Front Plant Sci 4:354

Baluška F, Mancuso S (2016) Vision in plants via plant-specific ocelli? Trends Plant Sci 21:727–730

Baluška F, Mancuso S (2017) Plant ocelli for visually guided plant behavior. Trends Plant Sci 22:5–6

Baluška F, Vitha S, Barlow PW, Volkmann D (1997) Rearrangements of F-actin arrays in growing cells of intact maize root apex tissues: a major developmental switch occurs in the postmitotic transition region. Eur J Cell Biol 72:113–121

Baluška F, Hlavačka A, Šamaj J, Palme K, Robinson DG, Matoh T, McCurdy DW, Menzel D, Volkmann D (2002) F-actin-dependent endocytosis of cell wall pectins in meristematic root cells: insights from brefeldin A-induced compartments. Plant Physiol 130:422–431

Baluška F, Šamaj J, Menzel D (2003) Polar transport of auxin: carrier-mediated flux across the plasma membrane or neurotransmitter-like secretion? Trends Cell Biol 13:282–285

Baluška F, Mancuso S, Volkmann D, Barlow PW (2004a) Root apices as plant command centres: the unique 'brain-like' status of the root apex transition zone. Biologia 59(Suppl. 13):9–17

Baluška F, Volkmann D, Barlow PW (2004b) Cell bodies in a cage. Nature 428:371

Baluška F, Volkmann D, Menzel D (2005a) Plant synapses: actin-based adhesion domains for cell-to-cell communication. Trends Plant Sci 10:106–111

Baluška F, Liners F, Hlavačka A, Schlicht M, Van Cutsem P, McCurdy D, Menzel D (2005b) Cell wall pectins and xyloglucans are internalized into dividing root cells and accumulate within cell plates during cytokinesis. Protoplasma 225:141–155

Baluška F, Barlow PW, Volkmann D, Mancuso S (2007) Gravity related paradoxes in plants: plant neurobiology provides the means for their resolution. In: Witzany G (ed) Biosemiotics in transdisciplinary context, Proceedings of the Gathering in Biosemiotics 6, Salzburg 2006. Umweb, Helsinky

Baluška F, Schlicht M, Volkmann D, Mancuso S (2008) Vesicular secretion of auxin: evidences and implications. Plant Signal Behav 3:254–256

Baluška F, Schlicht M, Wan Y-L, Burbach C, Volkmann D (2009a) Intracellular domains and polarity in root apices: from synaptic domains to plant neurobiology. Nova Acta Leopold 96:103–122

Baluška F, Mancuso S, Volkmann D, Barlow PW (2009b) The 'root-brain' hypothesis of Charles and Francis Darwin: revival after more than 125 years. Plant Signal Behav 4:1121–1127

Baluška F, Mancuso S, Volkmann D, Barlow PW (2010) Root apex transition zone: a signalling—response nexus in the root. Trends Plant Sci 15:402–408

Baluška F, Volkmann D, Menzel D, Barlow PW (2012) Strasburger's legacy to mitosis and cytokinesis and its relevance for the Cell Theory. Protoplasma 249:1151–1162

Baluška F, Yokawa K, Mancuso S, Baverstock K (2016) Understanding of anesthesia—why consciousness is essential for life and not based on genes. Commun Integr Biol 9:e1238118

Balzergue C, Dartevelle T, Godon C, Laugier E, Meisrimler C, Teulon JM, Creff A, Bissler M, Brouchoud C, Hagège A, Müller J, Chiarenza S, Javot H, Becuwe-Linka N, David P, Péret B, Delannoy E, Thibaud MC, Armengaud J, Abel S, Pellequer JL, Nussaume L, Desnos T (2017) Low phosphate activates STOP1-ALMT1 to rapidly inhibit root cell elongation. Nat Commun 8:15300

Baracchi D, Marples A, Jenkins AJ, Leitch AR, Chittka L (2017) Nicotine in floral nectar pharmacologically influences bumblebee learning of floral features. Sci Rep 7:1951

Barber CN, Huganir RL, Raben DM (2017) Phosphatidic acid-producing enzymes regulating the synaptic vesicle cycle: role for PLD? Adv Biol Regul (In press)

Barlow PW (1995) Gravity perception in plants—a multiplicity of systems derived by evolution. Plant Cell Environ 18:951–962

Barlow PW (2006) Charles Darwin and the plant root apex: closing a gap in living systems theory as applied to plants. In: Baluška F, Mancuso S, Volkmann D (eds) Communication in plants: neuronal aspect of plant life. Springer, Berlin, pp 37–51

Barlow SE, Wright GA, Ma C, Barberis M, Farrell IW, Marr EC, Brankin A, Pavlik BM, Stevenson PC (2017) Distasteful nectar deters floral robbery. Curr Biol 27:2552–2558

Barrada A, Montané MH, Robaglia C, Menand B (2015) Spatial regulation of root growth: placing the plant TOR pathway in a developmental perspective. Int J Mol Sci 16:19671–19697

Bateman RM (1991) Palaeoecology. In: Cleal CJ (ed) Plant fossils in geological investigation: the Palaeozoic. Ellis Horwood, Chichester, pp 34–116

Bateman RM, Crane PR, DiMichele WA, Kenrick P, Rowe NP, Speck T, Stein WE (1998) Early evolution of land plants: phylogeny, physiology and ecology of the primary terrestrial radiation. Annu Rev Ecol Syst 29:263–292

Bauer U, Di Giusto B, Skepper J, Grafe TU, Federle W (2012) With a flick of the lid: a novel trapping mechanism in *Nepenthes gracilis* pitcher plants. PLoS One 7:e38951

Bellot A, Guivernau B, Tajes M, Bosch-Morató M, Valls-Comamala V, Muñoz FJ (2014) The structure and function of actin cytoskeleton in mature glutamatergic dendritic spines. Brain Res 1573:1–16

Bender RL, Fekete ML, Klinkenberg PM, Hampton M, Bauer B, Malecha M, Lindgren K, Maki JA, Perera MA, Nikolau BJ, Carter CJ (2013) PIN6 is required for nectary auxin response and short stamen development. Plant J 74:893–890

Besserer A, Puech-Pagès V, Kiefer P, Gomez-Roldan V, Jauneau A, Roy S, Portais JC, Roux C, Bécard G, Séjalon-Delmas N (2006) Strigolactones stimulate arbuscular mycorrhizal fungi by activating mitochondria. PLoS Biol 4:e226

Bhatt MV, Khandelwal A, Dudley SA (2011) Kin recognition, not competitive interactions, predicts root allocation in young *Cakile edentula* seedling pairs. New Phytol 189:1135–1142

Białasek M, Górecka M, Mittler R, Karpiński S (2017) Evidence for the involvement of electrical, calcium and ROS signaling in the systemic regulation of non-photochemical quenching and photosynthesis. Plant Cell Physiol 58:207–215

Biedrzycki ML, Bais HP (2010) Kin recognition: another biological function for root secretions. Plant Signal Behav 5:401–402

Biedrzycki ML, Jilany TA, Dudley SA, Bais HP (2010) Root exudates mediate kin recognition in plants. Commun Integr Biol 3:28–35

Biegler R (2017) Insufficient evidence for habituation in *Mimosa pudica*. Response to Gagliano et al. (2014). Oecologia (In press)

Böhm J, Scherzer S, Krol E, Kreuzer I, von Meyer K, Lorey C, Mueller TD, Shabala L, Monte I, Solano R, Al-Rasheid KA, Rennenberg H, Shabala S, Neher E, Hedrich R (2016) The Venus flytrap *Dionaea muscipula* counts prey-induced action potentials to induce sodium uptake. Curr Biol 26:286–295

Bonhomme V, Pelloux-Prayer H, Jousselin E, Forterre Y, Labat JJ, Gaume L (2011) Slippery or sticky? Functional diversity in the trapping strategy of Nepenthes carnivorous plants. New Phytol 191:545–554

Brenner E, Stahlberg R, Mancuso S, Vivanco J, Baluška F, Van Volkenburgh E (2006) Plant neurobiology: an integrated view of plant signaling. Trends Plant Sci 11:413–419

Brenner ED, Stahlberg R, Mancuso S, Baluška F, Van Volkenburgh E (2007) Response to Alpi et al: plant neurobiology: the gain is more than the name. Trends Plant Sci 12:285–286

Broderbauer D, Weber A, Diaz A (2013) The design of trapping devices in pollination traps of the genus Arum (Araceae) is related to insect type. Bot J Linn Soc 172:385–397

Buell CR, Last RL (2010) Twenty-first century plant biology: impacts of the Arabidopsis genome on plant biology and agriculture. Plant Physiol 154:497–500

Bulgarelli D, Rott M, Schlaeppi K, van Themaat EVL, Ahmadinejad N, Assenza F, Rauf P, Huettel B, Reinhardt R, Schmelzer E, Peplies J, Gloeckner FO, Amann R, Eickhorst T, Schulze-Lefert P (2012) Revealing structure and assembly cues for Arabidopsis root-inhabiting bacterial microbiota. Nature 488:91–95

Burbach C, Markus K, Zhang Y, Schlicht M, Baluška F (2012) Photophobic behavior of maize roots. Plant Signal Behav 7:874–878

Busch W, Benfey PN (2010) Information processing without brains—the power of intercellular regulators in plants. Development 137:1215–1226

Calvo P (2017) What is it like to be a plant? J Conscious Stud 24:205–227

Calvo P, Friston K (2017) Predicting green: really radical (plant) predictive processing. J R Soc Interface 14:20170096

Calvo P, Keijzer FA (2011) Plants: adaptive behavior, root brains and minimal cognition. Adapt Behav 19:155–171

Calvo P, Baluška F, Sims A (2016) "Feature Detection" vs. "Predictive Coding" Models of plant behavior. Front Psychol 7:1505

Calvo P, Sahi VP, Trewavas A (2017) Are plants sentient? Plant Cell Environ 40:2858–2869

Cao Y, Tanaka K, Nguyen CT, Stacey G (2014) Extracellular ATP is a central signaling molecule in plant stress responses. Curr Opin Plant Biol 20:82–87

Carvalhais LC, Dennis PG, Badri DV, Kidd BN, Vivanco JM, Schenk PM (2015) Linking jasmonic acid signaling, root exudates, and rhizosphere microbiomes. Mol Plant-Microbe Interact 28:1049–1058

Cazzonelli CI, Vanstraelen M, Simon S, Yin K, Carron-Arthur A, Nisar N, Tarle G, Cuttriss AJ, Searle IR, Benkova E, Mathesius U, Masle J, Friml J, Pogson BJ (2013) Role of the Arabidopsis PIN6 auxin transporter in auxin homeostasis and auxin-mediated development. PLoS One 8: e70069

Chen R, Hilson P, Sedbrook J, Rosen E, Caspar T, Masson PH (1998) The *Arabidopsis thaliana* AGRAVITROPIC 1 gene encodes a component of the polar auxin transport efflux carrier. Proc Natl Acad Sci U S A 95:15112–15117

Chen BJ, During HJ, Anten NP (2012) Detect thy neighbor: identity recognition at the root level in plants. Plant Sci 195:157–167

Chiu SL, Diering GH, Ye B, Takamiya K, Chen CM, Jiang Y, Niranjan T, Schwartz CE, Wang T, Huganir RL (2017) GRASP1 regulates synaptic plasticity and learning through endosomal recycling of AMPA receptors. Neuron 93:1405–1419

Choi J, Tanaka K, Cao Y, Qi Y, Qiu J, Liang Y, Lee SY, Stacey G (2014a) Identification of a plant receptor for extracellular ATP. Science 343:290–294

Choi J, Tanaka K, Liang Y, Cao Y, Lee SY, Stacey G (2014b) Extracellular ATP, a danger signal, is recognized by DORN1 in Arabidopsis. Biochem J 463:429–437

Chowdhury MEK, Lim H, Bae H (2014) Update on the effects of sound wave on plants. Res Plant Dis 20:1–7

Contreras-Cornejo HA, López-Bucio JS, Méndez-Bravo A, Macías-Rodríguez L, Ramos-Vega M, Guevara-García ÁA, López-Bucio J (2015) Mitogen-activated protein Kinase 6 and ethylene and auxin signaling pathways are involved in Arabidopsis root-system architecture alterations by *Trichoderma atroviride*. Mol Plant-Microbe Interact 28:701–710

Couvillon MJ, Al Toufailia H, Butterfield TM, Schrell F, Ratnieks FLW, Schürch R (2015) Caffeinated forage tricks honeybees into increasing foraging and recruitment behaviors. Curr Biol 25:2815–2818

Cozzolino S, Widmer A (2005) Orchid diversity: an evolutionary consequence of deception? Trends Ecol Evol 20:487–494

Crepet WL, Niklas KJ (2009) Darwin's second 'abominable mystery': why are there so many angiosperm species? Am J Bot 96:366–381

Cvrčková F, Žárský V, Markoš A (2016) Plant studies may lead us to rethink the concept of behavior. Front Psychol 7:622

Darwin C (1880) Power of movements in plants. John Murray, London

Daspute AA, Sadhukhan A, Tokizawa M, Kobayashi Y, Panda SK, Koyama H (2017) Transcriptional regulation of aluminum-tolerance genes in higher plants: clarifying the underlying molecular mechanisms. Front Plant Sci 8:1358

De Luca PA, Vallejo-Marin M (2013) What's the 'buzz' about? The ecology and evolutionary significance of buzz-pollination. Curr Opin Plant Biol 16:429–435

de Wit M, Keuskamp DH, Bongers FJ, Hornitschek P, Gommers CMM, Reinen E, Martínez-Cerón C, Fankhauser C, Pierik R (2016) Integration of phytochrome and cryptochrome signals determines plant growth during competition for light. Curr Biol 26:3320–3326

Degenhardt J, Hiltpold I, Köllner TG, Frey M, Gierl A, Gershenzon J, Hibbard BE, Ellersieck MR, Turlings TC (2009) Restoring a maize root signal that attracts insect-killing nematodes to control a major pest. Proc Natl Acad Sci U S A 106:13213–13218

Dhonukshe P, Baluška F, Schlicht M, Hlavačka A, Šamaj J, Friml J, Gadella TWJ Jr (2006) Endocytosis of cell surface material mediates cell plate formation during plant cytokinesis. Dev Cell 10:137–150

Dhonukshe P, Grigoriev I, Fischer R, Tominaga M, Robinson DG, Hasek J, Paciorek T, Petrásek J, Seifertová D, Tejos R, Meisel LA, Zazímalová E, Gadella TW Jr, Stierhof YD, Ueda T, Oiwa K, Akhmanova A, Brock R, Spang A, Friml J (2008) Auxin transport inhibitors impair vesicle motility and actin cytoskeleton dynamics in diverse eukaryotes. Proc Natl Acad Sci U S A 105:4489–4494

Ditengou FA, Gomes D, Nziengui H, Kochersperger P, Lasok H, Medeiros V, Paponov IA, Nagy SK, Nádai TV, Mészáros T, Barnabás B, Ditengou BI, Rapp K, Qi L, Li X, Becker C, Li C, Dóczi R, Palme K (2017) Characterization of auxin transporter PIN6 plasma membrane targeting reveals a function for PIN6 in plant bolting. New Phytol. (In press)

Dittmer HJ (1937) A quantitative study of the roots and root hairs of a winter rye plant (*Secale cereale*). Am J Bot 24:417–420

Dolan L (1998) Pointing roots in the right direction: the role of auxin transport in response to gravity. Genes Dev 12:2091–2095

Dubos C, Huggins D, Grant GH, Knight MR, Campbell MM (2003) A role for glycine in the gating of plant NMDA-like receptors. Plant J 35:800–810

Eggenberger K, Sanyal P, Hundt S, Wadhwani P, Ulrich AS, Nick P (2017) Challenge integrity: the cell-penetrating peptide BP100 interferes with the auxin-actin oscillator. Plant Cell Physiol 58:71–85

Estabrook EM, Yoder JI (1998) Plant-plant communications: rhizosphere signaling between parasitic angiosperms and their hosts. Plant Physiol 116:1–7

Eysholdt-Derzsó E, Sauter M (2017) Root bending is antagonistically affected by hypoxia and ERF-mediated transcription via auxin signaling. Plant Physiol 175:412–423

Falik O, de Kroon H, Novoplansky A (2006) Physiologically-mediated self/non-self root discrimination in *Trifolium repens* has mixed effects on plant performance. Plant Signal Behav 1:116–121

Fan W, Lou HQ, Yang JL, Zheng SJ (2016) The roles of STOP1-like transcription factors in aluminum and proton tolerance. Plant Signal Behav 11:e1131371

Fasano J, Massa G, Gilroy S (2002) Ionic signaling in plant responses to gravity and touch. J Plant Growth Regul 21:71–88

Foo E, Yoneyama K, Hugill C, Quittenden LJ, Reid JB (2013) Strigolactones: internal and external signals in plant symbioses? Plant Signal Behav 8:e23168

Frederickson ME, Gordon DM (2007) The devil to pay: a cost of mutualism with *Myrmelachista schumanni* ants in 'devil's gardens' is increased herbivory on *Duroia hirsuta* trees. Proc Biol Sci 274:1117–1123

Frederickson ME, Greene MJ, Gordon DM (2005) 'Devil's gardens' bedevilled by ants. Nature 437:495–496

Friedman WE (2009) The meaning of Darwin's 'abominable mystery'. Am J Bot 96:5–21

Fuller S (2003) Kuhn versus Popper—the struggle for the soul of science. Columbia University Press, New York

Gagliano M (2013) Green symphonies: a call for studies on acoustic communication in plants. Behav Ecol 24:789–796

Gagliano M (2017) The mind of plants: thinking the unthinkable. Commun Integr Biol 10: e1288333

Gagliano M, Mancuso S, Robert D (2012) Towards understanding plant bioacoustics. Trends Plant Sci 17:323–325

Gagliano M, Vyazovskiy VV, Borbély AA, Grimonprez M, Depczynski M (2016) Learning by association in plants. Sci Rep 6:38427

Gagliano M, Grimonprez M, Depczynski M, Renton M (2017a) Tuned in: plant roots use sound to locate water. Oecologia 184:151–160

Gagliano M, Abramson CI, Depczynski M (2017b) Plants learn and remember: lets get used to it. Oecologia. (In press)

Galen C, Rabenold JJ, Liscum E (2007a) Functional ecology of a blue light photoreceptor: effects of phototropin-1 on root growth enhance drought tolerance in *Arabidopsis thaliana*. New Phytol 173:91–99

Galen C, Rabenold JJ, Liscum E (2007b) Light-sensing in roots. Plant Signal Behav 2:106–108

Galloway AF, Pedersen MJ, Merry B, Marcus SE, Blacker J, Benning LG, Field KJ, Knox JP (2017) Xyloglucan is released by plants and promotes soil particle aggregation. New Phytol. (In press)

Galvan-Ampudia CS, Testerink C (2011) Salt stress signals shape the plant root. Curr Opin Plant Biol 14:296–302

Galvan-Ampudia CS, Julkowska MM, Darwish E, Gandullo J, Korver RA, Brunoud G, Haring MA, Munnik T, Vernoux T, Testerink C (2013) Halotropism is a response of plant roots to avoid a saline environment. Curr Biol 23:2044–2050

Gardiner J (2012) Insights into plant consciousness from neuroscience, physics and mathematics: a role for quasicrystals? Plant Signal Behav 7:1049–1055

Gaskett AC, Winnick CG, Herberstein ME (2008) Orchid sexual deceit provokes ejaculation. Am Nat 171:E206–E212

Giehl RFH, von Wirén N (2014) Root nutrient foraging. Plant Physiol 166:509–517

Gilroy S, Białasek M, Suzuki N, Górecka M, Devireddy AR, Karpiński S, Mittler R (2016) ROS, calcium, and electric signals: key mediators of rapid systemic signaling in plants. Plant Physiol 171:1606–1615

Godfrey-Smith P (1996) Precis of complexity and the function of mind in nature. Adapt Behav 4:453–465

Golomb L, Abu-Abied M, Belausov E, Sadot E (2008) Different subcellular localizations and functions of Arabidopsis myosin VIII. BMC Plant Biol 8:3

González AP, Chrtek J, Dobrev PI, Dumalasová V, Fehrer J, Mráz P, Latzel V (2016) Stress-induced memory alters growth of clonal offspring of white clover (*Trifolium repens*). Am J Bot 103:1567–1574

González-Teuber M, Kaltenpoth M, Boland W (2014) Mutualistic ants as an indirect defence against leaf pathogens. New Phytol 202:640–650

Grasso DA, Pandolfi C, Bazihizina N, Nocentini D, Nepi M, Mancuso S (2015) Extrafloral-nectar-based partner manipulation in plant-ant relationships. AoB Plants 7:plv002

Grémiaux A, Yokawa K, Mancuso S, Baluška F (2014) Plant anesthesia supports similarities between animals and plants: Claude Bernard's forgotten studies. Plant Signal Behav 9:e27886

Gruntman M, Novoplansky A (2004) Physiologically mediated self/non-self discrimination in roots. Proc Natl Acad Sci U S A 101:3863–3867

Gu Y, Huganir RL (2016) Identification of the SNARE complex mediating the exocytosis of NMDA receptors. Proc Natl Acad Sci U S A 113:12280–12285

Gu Y, Chiu SL, Liu B, Wu PH, Delannoy M, Lin DT, Wirtz D, Huganir RL (2016) Differential vesicular sorting of AMPA and GABAA receptors. Proc Natl Acad Sci U S A 113:E922–E931

Gundel PE, Pierik R, Mommer L, Ballaré CL (2014) Competing neighbors: light perception and root function. Oecologia 176:1–10

Haberlandt T-H (1900) Über die Perception des geotropischen Reizes. Ber Dtsch Bot Ges 18:261–272

Hahn A, Firn R, Edelmann HG (2006) Interacting signal transduction chains in gravity-stimulated maize roots. Signal Transduct 6:449–455

Hahn A, Zimmermann R, Wanke D, Harter K, Edelmann HG (2008) The root cap determines ethylene-dependent growth and development in maize roots. Mol Plant 1:359–367

Hedrich R, Salvador-Recatalà V, Dreyer I (2016) Electrical wiring and long-distance plant communication. Trends Plant Sci 21:376–387

Henning T, Weigend M (2013a) Total control—pollen presentation and floral longevity in Loasaceae (blazing star family) are modulated by light, temperature and pollinator visitation rates. PLoS One 7:e41121

Henning T, Weigend M (2013b) Beautiful, complicated—and intelligent? Novel aspects of the thigmonastic stamen movement in Loasaceae. Plant Signal Behav 8:e24605

Hodge A (2009) Root decisions. Plant Cell Environ 32:628–640

Hoffland E, Findenegg GR, Nelemans JA (1989) Solubilization of rock phosphate by rape. II. Local root exudation of organic acids as a response to P-starvation. Plant Soil 113:161–165

Hotulainen P, Hoogenraad CC (2010) Actin in dendritic spines: connecting dynamics to function. J Cell Biol 189:619–629

Huang X, Maisch J, Nick P (2017) Sensory role of actin in auxin-dependent responses of tobacco BY-2. J Plant Physiol 218:6–15

Irwin RE, Cook D, Richardson LL, Manson JS, Gardner DR (2014) Secondary compounds in floral rewards of toxic rangeland plants: impacts on pollinators. J Agric Food Chem 62:7335–7344

Jásik J, Boggetti B, Baluška F, Volkmann D, Gensch T, Rutten T, Altmann T, Schmelzer E (2013) PIN2 turnover in Arabidopsis root epidermal cells explored by the photoconvertible protein Dendra2. PLoS One 8:e61403

Jásik J, Bokor B, Stuchlík S, Mičieta K, Turňa J, Schmelzer E (2016) Effects of auxins on PIN-FORMED2 (PIN2) dynamics are not mediated by inhibiting PIN2 endocytosis. Plant Physiol 172:1019–1031

Jersáková J, Johnson SD, Kindlmann P (2006) Mechanisms and evolution of deceptive pollination in orchids. Biol Rev Camb Philos Soc 81:219–235

Karban R (2008) Plant behavior and communication. Ecol Lett 11:727–739

Karban R (2015) Plant sensing and communication. University of Chicago Press, Chicago

Karpiński S, Szechyńska-Hebda M (2010) Secret life of plants: from memory to intelligence. Plant Signal Behav 5:1391–1394

Karpinski S, Reynolds H, Karpinska B, Wingsle G, Creissen G, Mullineaux P (1999) Systemic signaling and acclimation in response to excess excitation energy in Arabidopsis. Science 284:654–657

Kawa D, Julkowska MM, Sommerfeld HM, Ter Horst A, Haring MA, Testerink C (2016) Phosphate-dependent root system architecture responses to salt stress. Plant Physiol 172:690–706

Keijzer FA (2017) Evolutionary convergence and biologically embodied cognition. Interface Focus 7:20160123

Keijzer FA, van Duijn M, Lyon P (2013) What nervous systems do: early evolution, input–output, and the skin brain thesis. Adapt Behav 21:67–85

Kent Peters N, Long SR (1988) Alfalfa root exudates and compounds which promote or inhibit induction of *Rhizobium meliloti* nodulation genes. Plant Physiol 88:396–400

Kessler D, Baldwin IT (2007) Making sense of nectar scents: the effects of nectar secondary metabolites on floral visitors of *Nicotiana attenuata*. Plant J 49:840–354

Kessler D, Gase K, Baldwin IT (2008) Field experiments with transformed plants reveal the sense of floral scents. Science 321:1200–1202

Kessler D, Diezel C, Baldwin IT (2010) Changing pollinators as a means of escaping herbivores. Curr Biol 20:237–242

Knee EM, Gong FC, Gao M, Teplitski M, Jones AR, Foxworthy A, Mort AJ, Bauer WD (2001) Root mucilage from pea and its utilization by rhizosphere bacteria as a sole carbon source. Mol Plant-Microbe Interact 14:775–784

Kobayashi Y, Kobayashi Y, Sugimoto M, Lakshmanan V, Iuchi S, Kobayashi M, Bais HP, Koyama H (2013a) Characterization of the complex regulation of AtALMT1 expression in response to phytohormones and other inducers. Plant Physiol 162:732–740

Kobayashi Y, Lakshmanan V, Kobayashi Y, Asai M, Iuchi S, Kobayashi M, Bais HP, Koyama H (2013b) Overexpression of AtALMT1 in the *Arabidopsis thaliana* ecotype Columbia results in enhanced Al-activated malate excretion and beneficial bacterium recruitment. Plant Signal Behav 8:e25565

Koch H, Stevenson PC (2017) Do linden trees kill bees? Reviewing the causes of bee deaths on silver linden (*Tilia tomentosa*). Biol Lett 13:20170484

Köhler A, Pirk CW, Nicolson SW (2012) Honeybees and nectar nicotine: deterrence and reduced survival versus potential health benefits. J Insect Physiol 58:286–292

Koltai H (2015) Cellular events of strigolactone signalling and their crosstalk with auxin in roots. J Exp Bot 66:4855–4861

Koncz C (2006) Dedication: George P. Rédei—Arabidopsis geneticist and polymath. Plant Breed Rev 26:1–33

Konings H (1995) Gravitropism of roots: an evaluation of progress during the last three decades. Acta Bot Neerl 44:195–223

Kuhn TS (1962, 2012) The structure of scientific revolutions. University of Chicago Press, Chicago

Kumar AS, Bais HP (2012) Wired to the roots: impact of root-beneficial microbe interactions on aboveground plant physiology and protection. Plant Signal Behav 7:1598–1604

Kumar M, Pandya-Kumar N, Dam A, Haor H, Mayzlish-Gati E, Belausov E, Wininger S, Abu-Abied M, McErlean CS, Bromhead LJ, Prandi C, Kapulnik Y, Koltai H (2015) Arabidopsis response to low-phosphate conditions includes active changes in actin filaments and PIN2 polarization and is dependent on strigolactone signalling. J Exp Bot 66:1499–2510

Kutschera U, Niklas KJ (2009) Evolutionary plant physiology: Charles Darwin's forgotten synthesis. Naturwissenschaften 96:1339–1354

Lakshmanan V, Castaneda R, Rudrappa T, Bais HP (2013) Root transcriptome analysis of *Arabidopsis thaliana* exposed to beneficial *Bacillus subtilis* FB17 rhizobacteria revealed genes for bacterial recruitment and plant defense independent of malate efflux. Planta 238:657–668

Lapsansky ER, Milroy AM, Andales MJ, Vivanco JM (2016) Soil memory as a potential mechanism for encouraging sustainable plant health and productivity. Curr Opin Biotechnol 38:137–142

Lee HJ, Ha JH, Kim SG, Choi HK, Kim ZH, Han YJ, Kim JI, Oh Y, Fragoso V, Shin K, Hyeon T, Choi HG, Oh KH, Baldwin IT, Park CM (2016a) Stem-piped light activates phytochrome B to trigger light responses in *Arabidopsis thaliana* roots. Sci Signal 9:ra106

Lee HJ, Ha JH, Park CM (2016b) Underground roots monitor aboveground environment by sensing stem-piped light. Commun Integr Biol 9:e1261769

Lee HJ, Park YJ, Ha JH, Baldwin IT, Park CM (2017) Multiple routes of light signaling during root photomorphogenesis. Trends Plant Sci 22:803–812

Lei W, Myers KR, Rui Y, Hladyshau S, Tsygankov D, Zheng JQ (2017) Phosphoinositide-dependent enrichment of actin monomers in dendritic spines regulates synapse development and plasticity. J Cell Biol 216:2551–2564

Lev-Yadun S (2016a) Plants are not sitting ducks waiting for herbivores to eat them. Plant Signal Behav 11:e1179419

Lev-Yadun S (2016b) Does the whistling thorn acacia (*Acacia drepanolobium*) use auditory aposematism to deter mammalian herbivores? Plant Signal Behav 11:e1207035

Lev-Yadun S (2017) Do carrots outsmart rabbits? Trends Ecol Evol 32:227–228

Li X, Zhang W (2008) Salt-avoidance tropism in *Arabidopsis thaliana*. Plant Signal Behav 3:351–353

Liu Y, Lai N, Gao K, Chen F, Yuan L, Mi G (2013) Ammonium inhibits primary root growth by reducing the length of meristem and elongation zone and decreasing elemental expansion rate in the root apex in *Arabidopsis thaliana*. PLoS One 8:e61031

Liu JT, Hu LS, Liu YL, Chen RS, Cheng Z, Chen SJ, Amatore C, Huang WH, Huo KF (2014) Real-time monitoring of auxin vesicular exocytotic efflux from single plant protoplasts by amperometry at microelectrodes decorated with nanowires. Angew Chem Int Ed Eng 53:2643–2647

López-Bucio JS, Dubrovsky JG, Raya-González J, Ugartechea-Chirino Y, López-Bucio J, de Luna-Valdez LA, Ramos-Vega M, León P, Guevara-García AA (2014) *Arabidopsis thaliana* mitogen-activated protein kinase 6 is involved in seed formation and modulation of primary and lateral root development. J Exp Bot 65:169–183

López-Ráez JA, Shirasu K, Foo E (2017) Strigolactones in plant interactions with beneficial and detrimental organisms: the Yin and Yang. Trends Plant Sci 22:527–537

Luschnig C, Vert G (2014) The dynamics of plant plasma membrane proteins: PINs and beyond. Development 141:2924–2938

Luschnig C, Gaxiola RA, Grisafi P, Fink GR (1998) EIR1, a root-specific protein involved in auxin transport, is required for gravitropism in *Arabidopsis thaliana*. Genes Dev 12:2175–2187

Lyon P (2006) The biogenic approach to cognition. Cogn Process 7:11–29

Lyon P (2015) The cognitive cell: bacterial behavior reconsidered. Front Microbiol 6:264

Mancuso S, Marras AM, Volker M, Baluška F (2005) Non-invasive and continuous recordings of auxin fluxes in intact root apex with a carbon-nanotube-modified and self-referencing microelectrode. Anal Biochem 341:344–351

Mancuso S, Marras AM, Mugnai S, Schlicht M, Zarsky V, Li G, Song L, Hue HW, Baluška F (2007) Phospholipase Dζ2 drives vesicular secretion of auxin for its polar cell-cell transport in the transition zone of the root apex. Plant Signal Behav 2:240–244

Manoli A, Begheldo M, Genre A, Lanfranco L, Trevisan S, Quaggiotti S (2014) NO homeostasis is a key regulator of early nitrate perception and root elongation in maize. J Exp Bot 65:185–200

Manoli A, Trevisan S, Voigt B, Yokawa K, Baluška F, Quaggiotti S (2016) Nitric oxide-mediated maize root apex responses to nitrate are regulated by auxin and strigolactones. Front Plant Sci 6:1269

Marder M (2012) Plant intentionality and the phenomenological framework of plant intelligence. Plant Signal Behav 7:1365–1372

Marder M (2013) Plant intelligence and attention. Plant Signal Behav 8:e23902

Margulis L, Sagan D (1995) What is life? Simon & Schuster, New York

Masi E, Ciszak M, Stefano G, Renna L, Azzarello E, Pandolfi C, Mugnai S, Baluška F, Arecchi FT, Mancuso S (2009) Spatio-temporal dynamics of the electrical network activity in the root apex. Proc Natl Acad Sci U S A 106:4048–4053

Mayer VE, Frederickson ME, McKey D, Blatrix R (2014) Current issues in the evolutionary ecology of ant-plant symbioses. New Phytol 202:749–764

Mescher MC, De Moraes CM (2015) Role of plant sensory perception in plant-animal interactions. J Exp Bot 66:425–433

Mescher MC, Runyon JB, De Moraes CM (2006) Plant host finding by parasitic plants. A new perspective on plant-to-plant communication. Plant Signal Behav 1:284–286

Metlen KL, Aschehoug ET, Callaway RM (2009) Plant behavioural ecology: dynamic plasticity in secondary metabolites. Plant Cell Environ 32:641–653

Mettbach U, Strnad M, Mancuso S, Baluška F (2017) Immunogold-EM analysis reveal Brefeldin A-sensitive clusters of auxin in Arabidopsis root apex cells. Plant Signal Behav 10:e1327105

Michard E, Lima PT, Borges F, Silva AC, Portes MT, Carvalho JE, Gilliham M, Liu LH, Obermeyer G, Feijó JA (2011) Glutamate receptor-like genes form Ca^{2+} channels in pollen tubes and are regulated by pistil D-serine. Science 332:434–437

Mishra RC, Ghosh R, Bae H (2016) Plant acoustics: in the search of a sound mechanism for sound signaling in plants. J Exp Bot 67:4483–4494

Mo M, Yokawa K, Wan Y, Baluška F (2015) How and why do root apices sense light under the soil surface? Front Plant Sci 6:775

Moni A, Lee AY, Briggs WR, Han IS (2015) The blue light receptor Phototropin 1 suppresses lateral root growth by controlling cell elongation. Plant Biol 17:34–40

Monshausen GB, Zieschang HE, Sievers A (1996) Differential proton secretion in the apical elongation zone caused by gravistimulation is induced by a signal from the root cap. Plant Cell Environ 19:1408–1414

Mora-Macías J, Ojeda-Rivera JO, Gutiérrez-Alanís D, Yong-Villalobos L, Oropeza-Aburto A, Raya-González J, Jiménez-Domínguez G, Chávez-Calvillo G, Rellán-Álvarez R, Herrera-Estrella L (2017) Malate-dependent Fe accumulation is a critical checkpoint in the root developmental response to low phosphate. Proc Natl Acad Sci U S A 114:E3563–E3572

Morel J, Habib L, Plantureux S, Guckert A (1991) Influence of maize root mucilage on soil aggregate stability. Plant Soil 136:111–119

Müller A, Guan C, Gälweiler L, Tänzler P, Huijser P, Marchant A, Parry G, Bennett M, Wisman E, Palme K (1998) AtPIN2 defines a locus of Arabidopsis for root gravitropism control. EMBO J 17:6903–6911

Müller J, Beck M, Mettbach U, Komis G, Hause G, Menzel D, Šamaj J (2010) Arabidopsis MPK6 is involved in cell division plane control during early root development, and localizes to the pre-prophase band, phragmoplast, trans-golgi network and plasma membrane. Plant J 61:234–248

Murphy GP, Dudley SA (2009) Kin recognition: competition and cooperation in Impatiens (Balsaminaceae). Am J Bot 96:1990–1996

Nagawa S, Xu T, Lin D, Dhonukshe P, Zhang X, Friml J, Scheres B, Fu Y, Yang Z (2012) ROP GTPase-dependent actin microfilaments promote PIN1 polarization by localized inhibition of clathrin-dependent endocytosis. PLoS Biol 10:e1001299

Nemec B (1900) Über die Art der Wahrnehmung des Schwerkraftreizes bei den Pflanzen. Ber Dtsch Bot Ges 18:241–245

Nemec B (1902) Die Perception des Schwerkraftreizes bei den Pflanzen. Ber Dtsch Bot Ges 18:339–355

Nepi M (2014) Beyond nectar sweetness: the hidden ecological role of non-protein amino acids in nectar. J Ecol 102:108–115

Nepi M (2017) New perspectives in nectar evolution and ecology: simple alimentary reward or a complex multiorganism interaction? Acta Agrobot 70 (In press)

Ni M, Zhang L, Shi YF, Wang C, Lu Y, Pan J, Liu JZ (2017) Excessive cellular S-nitrosothiol impairs endocytosis of auxin efflux transporter PIN2. Front Plant Sci 8:1988

Nick P (2010) Probing the actin-auxin oscillator. Plant Signal Behav 5:94–98

Nisar N, Cuttriss AJ, Pogson BJ, Cazzonelli CI (2014) The promoter of the Arabidopsis PIN6 auxin transporter enabled strong expression in the vasculature of roots, leaves, floral stems and reproductive organs. Plant Signal Behav 9:e27898

Nonaka A, Toyoda T, Miura Y, Hitora-Imamura N, Naka M, Eguchi M, Yamaguchi S, Ikegaya Y, Matsuki N, Nomura H (2014) Synaptic plasticity associated with a memory engram in the basolateral amygdala. J Neurosci 34:9305–9309

Novoplansky A (2009) Picking battles wisely: plant behaviour under competition. Plant Cell Environ 32:726–741

Oelschlägel B, Gorb S, Wanke S, Neinhuis C (2009) Structure and biomechanics of trapping flower trichomes and their role in the pollination biology of Aristolochia plants (Aristolochiaceae). New Phytol 184:988–1002

Orrock J, Connolly B, Kitchen A (2017) Induced defences in plants reduce herbivory by increasing cannibalism. Nat Ecol Evol 1:1205–1207

Ortiz-Ramírez C, Michard E, Simon AA, Damineli DSC, Hernández-Coronado M, Becker JD, Feijó JA (2017) GLUTAMATE RECEPTOR-LIKE channels are essential for chemotaxis and reproduction in mosses. Nature 549:91–95

Pandya-Kumar N, Shema R, Kumar M, Mayzlish-Gati E, Levy D, Zemach H, Belausov E, Wininger S, Abu-Abied M, Kapulnik Y, Koltai H (2014) Strigolactone analog GR24 triggers changes in PIN2 polarity, vesicle trafficking and actin filament architecture. New Phytol 202:1184–1196

Pantazopoulou CK, Bongers FJ, Küpers JJ, Reinen E, Das D, Evers JB, Anten NPR, Pierik R (2017) Neighbor detection at the leaf tip adaptively regulates upward leaf movement through spatial auxin dynamics. Proc Natl Acad Sci U S A 114:7450–7455

Pavlovič A, Jakšová J, Novák O (2017) Triggering a false alarm: wounding mimics prey capture in the carnivorous Venus flytrap (*Dionaea muscipula*). New Phytol 216:927–938

Perrin RM, Young LS, Murthy UMN, Harrison BR, Wang Y, Will JL, Masson PH (2005) Gravity signal transduction in primary roots. Ann Bot 96:737–743

Phillips RD, Scaccabarozzi D, Retter BA, Hayes C, Brown GR, Dixon KW, Peakall R (2014) Caught in the act: pollination of sexually deceptive trap-flowers by fungus gnats in Pterostylis (Orchidaceae). Ann Bot 113:629–641

Pierik R, de Wit M (2014) Shade avoidance: phytochrome signalling and other aboveground neighbour detection cues. J Exp Bot 65:2815–2824

Pierik R, Testerink C (2014) The art of being flexible: how to escape from shade, salt, and drought. Plant Physiol 166:5–22

Pierik R, De Wit M, Voesenek LACJ (2011) Growth-mediated stress escape: convergence of signal transduction pathways activated upon exposure to two different environmental stresses. New Phytol 189:122–134

Plieth C, Trewavas A (2002) Reorientation of seedlings in the Earth's gravitational field induces cytosolic calcium transients. Plant Physiol 129:786–796

Popper K (1994) All life is problem solving. Routledge, Taylor and Francis Group, Abingdon

Price MB, Jelesko J, Okumoto S (2012) Glutamate receptor homologs in plants: functions and evolutionary origins. Front Plant Sci 3:235

Pyke GH (2016) Floral nectar: pollinator attraction or manipulation? Trends Ecol Evol 31:339–341

Qu Y, Wang Q, Guo J, Wang P, Song P, Jia Q, Zhang X, Kudla J, Zhang W, Zhang Q (2017) Peroxisomal CuAOζ and its product H_2O_2 regulate the distribution of auxin and IBA-dependent lateral root development in Arabidopsis. J Exp Bot 68:4851–4867

Raghothama KG (1999) Phosphate acquisition. Annu Rev Plant Physiol Plant Mol Biol 50:665–693

Ramesh SA, Tyerman SD, Xu B, Bose J, Kaur S, Conn V, Domingos P, Ullah S, Wege S, Shabala S, Feijó JA, Ryan PR, Gilliham M (2015) GABA signalling modulates plant growth by directly regulating the activity of plant-specific anion transporters. Nat Commun 6:7879

Ramesh SA, Tyerman SD, Gilliham M, Xu B (2017) γ-Aminobutyric acid (GABA) signalling in plants. Cell Mol Life Sci 74:1577–1603

Rasmann S, Köllner TG, Degenhardt J, Hiltpold I, Toepfer S, Kuhlmann U, Gershenzon J, Turlings TCJ (2005) Recruitment of entomopathogenic nematodes by insect-damaged maize roots. Nature 434:732–737

Retzer K, Lacek J, Skokan R, Del Genio CI, Vosolsobě S, Laňková M, Malínská K, Konstantinova N, Zažímalová E, Napier RM, Petrášek J, Luschnig C (2017) Evolutionary conserved cysteines function as cis-acting regulators of Arabidopsis PIN-FORMED 2 distribution. Int J Mol Sci 18:E2274

Rodrigo-Moreno A, Bazihizina N, Azzarello E, Masi E, Tran D, Bouteau F, Baluška F, Mancuso S (2017) Root phonotropism: early signalling events following sound perception in Arabidopsis roots. Plant Sci 264:9–15

Rudrappa T, Czymmek KJ, Paré PW, Bais HP (2008) Root-secreted malic acid recruits beneficial soil bacteria. Plant Physiol 148:1547–1556

Rumpel S, LeDoux J, Zador A, Malinow R (2005) Postsynaptic receptor trafficking underlying a form of associative learning. Science 308:83–88

Runyon JB, Mescher MC, De Moraes CM (2006) Volatile chemical cues guide host location and host selection by parasitic plants. Science 313:1964–2967

Šamaj J, Read ND, Volkmann D, Menzel D, Baluška F (2005) The endocytic network in plants. Trends Cell Biol 15:425–433

Sattarzadeh A, Franzen R, Schmelzer E (2008) The Arabidopsis class VIII myosin ATM2 is involved in endocytosis. Cell Motil Cytoskeleton 65:457–468

Schiestl FP (2005) On the success of a swindle: pollination by deception in orchids. Naturwissenschaften 92:255–264

Schiestl FP (2010) Pollination: sexual mimicry abounds. Curr Biol 20:R1020–R1022

Schlicht M, Strnad M, Scanlon MJ, Mancuso S, Hochholdinger F, Palme K, Volkmann D, Menzel D, Baluška F (2006) Auxin immunolocalization implicates vesicular neurotransmitter-like mode of polar auxin transport in root apices. Plant Signal Behav 1:122–133

Schlicht M, Šamajová O, Schachtschabel D, Mancuso S, Menzel D, Boland W, Baluška F (2008) Dorenone blocks polarized tip-growth of root hairs by interfering with the PIN2-mediated auxin transport network in the root apex. Plant J 55:709–717

Sekereš J, Pleskot R, Pejchar P, Žárský V, Potocký M (2015) The song of lipids and proteins: dynamic lipid-protein interfaces in the regulation of plant cell polarity at different scales. J Exp Bot 66:1587–1598

Sibaoka T (1991) Rapid plant movements triggered by action potentials. Bot Mag Tokyo 104:73–95

Sievers A, Behrens H, Buckhout T, Gradmann D (1984) Can a Ca^{2+} pump in the endoplasmic reticulum of the Lepidium root be the trigger for rapid changes in membrane potential after gravistimulation? J Plant Physiol 114:195–200

Simon S, Skůpa P, Viaene T, Zwiewka M, Tejos R, Klíma P, Čarná M, Rolčík J, De Rycke R, Moreno I, Dobrev PI, Orellana A, Zažímalová E, Friml J (2016) PIN6 auxin transporter at endoplasmic reticulum and plasma membrane mediates auxin homeostasis and organogenesis in Arabidopsis. New Phytol 211:65–74

Sivaguru M, Baluška F, Volkmann D, Felle HH, Horst WJ (1999) Impacts of aluminum on the cytoskeleton of the maize root apex. Short-term effects on the distal part of the transition zone. Plant Physiol 119:1073–1082

Sivaguru M, Pike S, Gassmann W, Baskin TI (2003) Aluminum rapidly depolymerizes cortical microtubules and depolarizes the plasma membrane: evidence that these responses are mediated by a glutamate receptor. Plant Cell Physiol 44:667–675

Smékalová V, Luptovčiak I, Komis G, Šamajová O, Ovečka M, Doskočilová A, Takáč T, Vadovič P, Novák O, Pechan T, Ziemann A, Košútová P, Šamaj J (2014) Involvement of YODA and mitogen activated protein kinase 6 in Arabidopsis post-embryogenic root development through auxin up-regulation and cell division plane orientation. New Phytol 203:1175–1193

Stahlberg R, Stephens NR, Cleland RE, Van Volkenburgh E (2006) Shade-induced action potentials in *Helianthus annuus* L. originate primarily from the epicotyl. Plant Signal Behav 1:15–22

Stolarz M, Dziubinska H (2017) Osmotic and salt stresses modulate spontaneous and glutamate-induced action potentials and distinguish between growth and circumnutation in *Helianthus annuus* seedlings. Front Plant Sci 8:1766

Sun F, Zhang W, Hu H, Li B, Wang Y, Zhao Y, Li K, Liu M, Li X (2008) Salt modulates gravity signaling pathway to regulate growth direction of primary roots in Arabidopsis. Plant Physiol 146:178–188

Sun J, Chen Q, Qi L, Jiang H, Li S, Xu Y, Liu F, Zhou W, Pan J, Li X, Palme K, Li C (2011) Jasmonate modulates endocytosis and plasma membrane accumulation of the Arabidopsis PIN2 protein. New Phytol 191:360–375

Suzuki H, Yokawa K, Nakano S, Yoshida Y, Fabrissin I, Okamoto T, Baluška F, Koshiba T (2016) Root cap-dependent gravitropic U-turn of maize root requires light-induced auxin biosynthesis via the YUC pathway in the root apex. J Exp Bot 67:4581–4591

Svistoonoff S, Creff A, Reymond M, Sigoillot-Claude C, Ricaud L, Blanchet A, Nussaume L, Desnos T (2007) Root tip contact with low-phosphate media reprograms plant root architecture. Nat Genet 39:792–796

Sweeney C, Lakshmanan V, Bais HP (2017) Interplant above-ground signaling prompts upregulation of auxin promoter and malate transporter as part of defensive response in the neighboring plants. Front Plant Sci 8:595

Szczuka A, Korczyńska J, Wnuk A, Symonowicz B, Gonzalez Szwacka A, Mazurkiewicz P, Kostowski W, Godzińska EJ (2013) The effects of serotonin, dopamine, octopamine and tyramine on behavior of workers of the ant *Formica polyctena* during dyadic aggression tests. Acta Neurobiol Exp. (Wars) 73:495–520

Szechyńska-Hebda M, Kruk J, Górecka M, Karpińska B, Karpiński S (2010) Evidence for light wavelength-specific photoelectrophysiological signaling and memory of excess light episodes in Arabidopsis. Plant Cell 22:2201–2218

Szechyńska-Hebda M, Lewandowska M, Karpiński S (2017) Electrical signaling, photosynthesis and systemic acquired acclimation. Front Physiol 8:684

Takáč T, Vadovič P, Pechan T, Luptovčiak I, Šamajová O, Šamaj J (2016) Comparative proteomic study of Arabidopsis mutants *mpk4* and *mpk6*. Sci Rep 6:28306

Tanaka K, Choi J, Cao Y, Stacey G (2014) Extracellular ATP acts as a damage-associated molecular pattern (DAMP) signal in plants. Front Plant Sci 5:446

Tononi G, Cirelli C (2014) Sleep and the price of plasticity: from synaptic and cellular homeostasis to memory consolidation and integration. Neuron 81:12–34

Trevisan S, Manoli A, Quaggiotti S (2014) NO signaling is a key component of the root growth response to nitrate in *Zea mays* L. Plant Signal Behav 9:e28290

Trevisan S, Manoli A, Ravazzolo L, Botton A, Pivato M, Masi A, Quaggiotti S (2015) Nitrate sensing by the maize root apex transition zone: a merged transcriptomic and proteomic survey. J Exp Bot 66:3699–3715

Trewavas A (2005) Plant intelligence. Naturwissenschaften 92:401–413

Trewavas A (2007) Response to Alpi et al.: Plant neurobiology—all metaphors have value. Trends Plant Sci 12:231–233

Trewavas AJ (2014) Plan behaviour and intelligence. Oxford University Press, Oxford

Trewavas A (2016) Intelligence, cognition, and language of green plants. Front Psychol 7:588

Trewavas A (2017) The foundations of plant intelligence. Interface Focus 7:20160098

Trewavas A, Baluška F (2011) The ubiquity of consciousness. EMBO Rep 12:1221–1225

Tripathi D, Zhang T, Koo AJ, Stacey G, Tanaka K (2017) Extracellular ATP acts on jasmonate signaling to reinforce plant defense. Plant Physiol (In press)

Tsai HH, Schmidt W (2017) Mobilization of iron by plant-borne coumarins. Trends Plant Sci 22:538–548

Tu-Sekine B, Goldschmidt H, Raben DM (2015) Diacylglycerol, phosphatidic acid, and their metabolic enzymes in synaptic vesicle recycling. Adv Biol Regul 57:147–152

Utsuno K, Shikanai T, Yamada Y, Hashimoto T (1998) AGR, an Agravitropic locus of *Arabidopsis thaliana*, encodes a novel, membrane-protein family member. Plant Cell Physiol 39:1111–1118

Valenzuela CE, Acevedo-Acevedo O, Miranda GS, Vergara-Barros P, Holuigue L, Figueroa CR, Figueroa PM (2016) Salt stress response triggers activation of the jasmonate signaling pathway leading to inhibition of cell elongation in Arabidopsis primary root. J Exp Bot 67:4209–4220

Vallverdú J, Castro O, Mayne R, Talanov M, Levin M, Baluška B, Gunji Y, Dussutour A, Zenil H, Adamatzky A (2018) Slime mould: the fundamental mechanisms of biological cognition. Biosystems 165:57–70

van Duijn M (2017) Phylogenetic origins of biological cognition: convergent patterns in the early evolution of learning. Interface Focus 7:20160158

Van Hoven W (1991) Mortalities in kudu (*Tragelaphus strepsiceros*) populations related to chemical defence in trees. J Afr Zool 105:141–146

Van Loon LC (2016) The intelligent behavior of plants. Trends Plant Sci 21:286–294

Volkmann D, Mori T, Tirlapur UK, König K, Fujiwara T, Kendrick-Jones J, Baluška F (2003) Unconventional myosins of the plant-specific class VIII: endocytosis, cytokinesis, plasmodesmata/pit-fields, and cell-to-cell coupling. Cell Biol Int 27:289–291

Walker TS, Bais HP, Grotewold E, Vivanco JM (2003) Root exudation and rhizosphere biology. Plant Physiol 132:44–51

Wan Y, Jasik J, Wang L, Hao H, Volkmann D, Menzel D, Mancuso S, Baluška F, Lin J (2012) The signal transducer NPH3 integrates the phototropin1 photosensor with PIN2-based polar auxin transport in Arabidopsis root phototropism. Plant Cell 24:551–565

Waters MT, Gutjahr C, Bennett T, Nelson DC (2017) Strigolactone signaling and evolution. Annu Rev Plant Biol 68:291–322

Watt M, McCully ME, Canny MJ (1994) Formation and stabilization of rhizosheaths of *Zea mays* L.: effect of soil water content. Plant Physiol 106:179–186

Weiland M, Mancuso S, Baluška F (2016) Signalling via glutamate and GLRs in *Arabidopsis thaliana*. Funct Plant Biol 43:1–25

Wittmann M, Queisser G, Eder A, Wiegert JS, Bengtson CP, Hellwig A, Wittum G, Bading H (2009) Synaptic activity induces dramatic changes in the geometry of the cell nucleus: interplay between nuclear structure, histone H3 phosphorylation, and nuclear calcium signaling. J Neurosci 29:14687–14700

Wolverton C, Ishikawa H, Evans M (2002) The kinetics of root gravitropism: dual motor and sensors. J Plant Growth Regul 21:102–112

Xu W, Jia L, Baluška F, Ding G, Shi W, Ye N, Zhang J (2012) PIN2 is required for the adaptation of Arabidopsis roots to alkaline stress by modulating proton secretion. J Exp Bot 63:6105–6114

Yan S, Zhang T, Dong S, McLamore ES, Wang N, Shan X, Shen Y, Wan Y (2016) MeJA affects root growth by modulation of transmembrane auxin flux in the transition zone. J Plant Growth Regul 35:256–265

Yan S, Jiao C, McLamore ES, Wang N, Yao H, Shen Y (2017) Insect herbivory of leaves affects the auxin flux along root apices in *Arabidopsis thaliana*. J Plant Growth Regul 36:846–854

Yang ZB, He C, Ma Y, Herde M, Ding Z (2017) Jasmonic acid enhances Al-induced root growth inhibition. Plant Physiol 173:1420–1433

Yasuda R (2017) Biophysics of biochemical signaling in dendritic spines: implications in synaptic plasticity. Biophys J 113:2152–2159

Yokawa K, Kagenishi T, Kawano T, Mancuso S, Baluška F (2011) Illumination of Arabidopsis roots induces immediate burst of ROS production. Plant Signal Behav 6:1457–1461

Yokawa K, Fassano R, Kagenishi T, Baluška F (2014) Light as stress factor to plant roots – case of root halotropism. Front Plant Sci 5:718

Yokawa K, Baluška F (2015) *C. elegans* and *Arabidopsis thaliana* show similar behavior: ROS induce escape tropisms both in illuminated nematodes and roots. Plant Signal Behav 10: e1073870

Yokawa K, Baluška F (2016) The TOR complex: an emergency switch for root behavior. Plant Cell Physiol 57:14–18

Yokawa K, Kagenishi T, Pavlovič A, Gall S, Weiland M, Mancuso S, Baluška F (2018) Anesthetics stop diverse plant organ movements, affect endocytic vesicle recycling, ROS homeostasis, and block action potentials in Venus Flytraps. Ann Bot (In press)

Žárský V (2015) Signal transduction: GABA receptor found in plants. Nat Plants 1:15115

Zhu J, Geisler M (2015) Keeping it all together: auxin-actin crosstalk in plant development. J Exp Bot 66:4983–4998

Zhu J, Bailly A, Zwiewka M, Sovero V, Di Donato M, Ge P, Oehri J, Aryal B, Hao P, Linnert M, Burgardt NI, Lücke C, Weiwad M, Michel M, Weiergräber OH, Pollmann S, Azzarello E, Mancuso S, Ferro N, Fukao Y, Hoffmann C, Wedlich-Söldner R, Friml J, Thomas C, Geisler M (2017a) TWISTED DWARF1 mediates the action of auxin transport inhibitors on actin cytoskeleton dynamics. Plant Cell 28:930–948

Zhu R, Dong X, Hao W, Gao W, Zhang W, Xia S, Liu T, Shang Z (2017b) Heterotrimeric G protein-regulated Ca^{2+} influx and PIN2 asymmetric distribution are involved in *Arabidopsis thaliana* roots' avoidance response to extracellular ATP. Front Plant Sci 8:1522

Zou N, Li B, Chen H, Su Y, Kronzucker HJ, Xiong L, Baluška F, Shi W (2013) GSA-1/ARG1 protects root gravitropism in Arabidopsis under ammonium stress. New Phytol 200:97–111

Role of Epigenetics in Transgenerational Changes: Genome Stability in Response to Plant Stress

Igor Kovalchuk

Abstract Mitotic or meiotic cell reproduction requires DNA replication. This mechanism ensures a faithful duplication of the exact genetic code. Whereas such approach is essential when organisms live in the stable environment, however, it does not allow adaptation to the contrasting environment because mutations are very rare and random. Fortunately, organisms can mount a quick epigenetic response including changes in DNA methylation, histone modifications, and the differential expression of various noncoding RNAs. These modifications do not change the genetic code, but rather allow cells and organisms to respond to the environment in a flexible and efficient manner. Epigenetic modifications are essential for the response to the environment at both somatic and transgenerational levels. The latter is especially important for the immediate plant survival and for the long-term adaptation to adverse conditions. In this chapter, we discuss various epigenetic mechanisms of regulation of genome stability, especially those ones that contribute to transgenerational changes.

Abbreviations

CAF-1	Chromatin assembly factor 1
DCL	Dicer-like
DDM1	DECREASED DNA METHYLATION1
diRNAs	DSB-induced ncRNAs
DSB	Double-strand break
HATs	Histone acetyltransferases
HDACs	Histone deacetylases
HP1	HETEROCHROMATIN PROTEIN1
HR	Homologous recombination
KDMs	Histone demethylases
MBDs	Methyl-CpG-binding domain proteins

I. Kovalchuk (✉)
Department of Biological Sciences, University of Lethbridge, Lethbridge, AB, Canada
e-mail: Igor.kovalchuk@uleth.ca

MMS   Methyl methanesulfonate
MOM1   MAINTENANCE OF METHYLATION 1
ncRNAs   Noncoding RNAs
NHEJ   Non-homologous end-joining
ORMV   Oilseed rape mosaic virus
RdDM   RNA-direct DNA methylation
smRNA   Small RNA
TMV   Tobacco mosaic virus

1 Introduction

The genetic material of every cell is typically faithfully reproduced through the replication process upon every cell division. Therefore, the genetic inheritance is based on the generation of identical copies of the genetic material. In contrast, the epigenetic inheritance represents a more flexible mechanism of control of gene expression and inheritance of old traits. At the same time, the epigenetic regulation leaves room for the appearance of new traits in response to the changing environmental conditions. Would it still be important to preserve the integrity of DNA if downstream mechanisms could potentially mitigate these changes? The epigenetic mechanisms are reversible, but they are more complex for inheritance compared to the genetic mechanisms. The preservation of master key—a genetic code—allows organisms to maintain the same developmental program and the same stable response to the constant environment to which these organisms (and species) have adapted to.

Plants have to maintain genome stability at every stage of development. DNA is under a constant threat from internal and external factors. An active metabolism, including photosynthesis and cellular respiration, poses a continuous challenge for plant genomes. These types of internal stresses produce free radicals that damage DNA directly or activate various signaling pathways potentially leading to secondary changes in genome stability. In addition, radicals oxidize lipids and proteins, thus altering their capacity to maintain genome stability.

The sedentary nature of plants makes them more vulnerable to the changing environment because plants cannot run away and avoid stress. External environmental stresses include chemical and physical stresses as well as changes in light intensity, temperature fluctuations, water and nutrient availability, wind, and other mechanical stimuli as well as biotic stresses in the form of various pathogens. See Madlung and Comai (Madlung and Comai 2004) for the details on various stresses that plants are exposed to in their environment.

Stress survival includes the entire battery of responses that are present in plants encoded in genetic and epigenetic mechanisms. At the same time, plants are also ableto develop new adaptive mechanisms that allow them to better survive new stress conditions. To cope with stress, plants use mechanisms of tolerance and resistance (Boyko and Kovalchuk 2011). These mechanisms are in part dependent on adaptive metabolic changes in somatic cells and heritable transgenerational changes (Chinnusamy and Zhu 2009). Plants have also the ability to maintain genome stability in the ever-changing growth environment (Dassler et al. 2008). This is in part due to the fact that plants possess additional copies of various DNA repair genes that often have redundant functions (Singh et al. 2010).

Genome integrity is maintained through various mechanisms, including the maintenance of nuclear architecture, the preservation of proper chromatin structure, and the utilization of different DNA repair pathways. DNA damage repair is controlled at multiple levels, including scanning of double-stranded DNA, the identification of damage, the global or local relaxation of chromatin, the recruitment of the repairsome, actual steps of DNA repair, the proofreading of newly added DNA sequences, and the reestablishment of similar or different epigenetic factors, including changes in DNA methylation and histone modifications (Tuteja et al. 2001). The degree of DNA damage and its specific location can perhaps be regulated at the level of different chromatin compaction; it is plausible to think that genome stability of a given chromosomal region can be relaxed not only by choosing different DNA repair pathways but also by introducing or removing various epigenetic modifications (Downey and Durocher 2006).

The control over the efficiency of DNA repair and the activity of mechanisms preserving genome stability is thus regulated by a variety of genetic and epigenetic factors (Downey and Durocher 2006). Whereas different DNA repair pathways are described elsewhere (Liu et al. 2006; Lombard et al. 2005), in this chapter, we will describe the epigenetic mechanisms that can regulate the mechanisms of genome stability, with a special attention being given to transgenerational effects.

2 The Choice of the DNA Repair Pathway may Influence Genome Stability

Strand breaks are dangerous DNA lesions, with double-strand break (DSB) being most deleterious. Two major repair pathways, non-homologous end-joining (NHEJ) and homologous recombination (HR), are involved in the repair of DSBs (Lieber 2010). NHEJ repair in most cases involves a direct rejoining of break ends; it either does not require any homology between the interacting DNA molecules or requires overlapping of a few nucleotides. In the cases when rejoining via a direct ligation process is impossible, Ku70/Ku80 proteins search for microhomology aligning one or several complementary bases to direct repeats, thus leading to the deletion of DNA between such repeats. As a result, NHEJ repair is mostly inaccurate and may result in large deletions spanning thousands of nucleotides (Lieber 2010). On the other side, the HR repair mechanism is quite accurate; it requires an extensive sequence homology and the presence of a repair template such as a sister chromatid or a homologous chromosome. Still when a template with imperfect homology is used, HR may result in gene conversion leading to a loss of heterozygosity. In some cases, when DNA regions containing multiple repeats are used, HR can lead to gene translocations, duplications, and deletion events as large as a chromosome arm. To summarize, NHEJ appears to be an error-prone repair mechanism of immediate response, whereas HR represents a rather slow but relatively error-free process (Chiruvella et al. 2013). A more detailed information about types of NHEJ and HR repairs in plants can be found elsewhere (Steinert et al. 2016).

The balance between NHEJ and HR occurrence is tightly regulated and depends on such factors as the availability of repair templates, the phase of a cell cycle, the rate of cell proliferation, and even a specific function of a given cell type [reviewed in (Shrivastav et al. 2008)]. NHEJ is the predominant DNA repair pathway mostly because it does not require a template and can deal with simple strand breaks in a fast and efficient manner; cells use it more often during the G1 phase of a cell cycle. In comparison, the HR pathway is relatively rarely used, perhaps in less than 1% of all cases of strand break repair, although it is more active (in relative terms, since it is still infrequent) during the S and G2 phases when sister chromatids are formed (Shibata et al. 2011). Thus, HR may play an essential role in the actively dividing cells, mostly during an early development of an organism. Plant tissues with a higher ploidy level seem to use HR less frequently; leaves of older plants that tend to have a larger number of genomes per single cell (due to endoreduplication) have a lower frequency of homologous recombination when prorated to a single genome (Boyko et al. 2006b). It also seems that the genome size has also some effects on the frequency of the use of HR repair as it may be more difficult to find a homologous template in a larger genome (Tiley and Burleigh 2015; Langley et al. 1988).

Also, the chromatin structure in large complex genomes may also contribute to an additional difficulty for HR to occur more frequently. Other repair mechanisms may also be altered by the chromatin structure, and this is reflected by the fact that the mutation frequency varies dramatically in different cell types and even during different developmental stages of an organism. Next, we attempted to summarize what is known about the epigenetic control over the stability of highly repetitive genomes in plants.

3 Epigenetic Mechanisms of Regulation of Plant Genome Stability

Unlike other organisms that can use the escape and avoidance mechanisms in response to stress, plants can only minimize the damage by the mechanisms of acclimation and adaptation. Plants are able to mount short-term and long-term defense responses to combat stress. Natural stresses are often chronic in nature as a very few of them can be considered acute. Stresses often result in changes in homeostasis as a result of both stresses itself (e.g., a chemical stress) and a plant response.

Rapid alterations in homeostasis, including massive changes in the number and amount of produced metabolites, are mechanisms through which plants respond to stresses. These changes in homeostasis include immediate responses that involve changes in gene expression leading to changes in RNA and protein levels, the synthesis and re-compartmentalization of various metabolites, the balance of salt concentration, pH, hormones, and many other events. Although these responses are critical for plant survival, their description is not the focus of this chapter; all the necessary information can be found elsewhere, including reviews by (Shinozaki et al. 2003; Sung and Amasino 2004).

Many events of response to stress are regulated through epigenetic mechanisms, including the repositioning of histone and nonhistone chromatin binding proteins in the nuclear matrix, changes in DNA methylation and histone modifications, the differential expression of ncRNAs, and ncRNA-mediated degradation of mRNA, among the others. All these mechanisms are important for the immediate survival of plants and for a long-term response. The epigenetic mechanisms involved in the regulation of genome stability in plants in response to stress are discussed below.

3.1 The Role of Chromatin Structure in Response to Stress and Genome Stability

The transcriptional activation or repression in response to stress is initiated at the level of chromatin and represent broadly either chromatin condensation or decondensation. Both processes involve the activity of ATP-dependent remodeling complexes, covalent modifications of histones, nucleosome eviction, histone variants, and/or changes in cytosine methylation. Moreover, noncoding small RNAs (smRNAs) contribute to these processes by altering the chromatin structure in a sequence-specific manner.

The condensed chromatin prevents the processes of transcription, replication, and DNA repair from occurring, and specific modifications of histones are required for DNA unpacking. The removal of nucleosomes from the damaged DNA allows a DNA repair protein to fix the damage, but it also makes DNA more vulnerable for an additional damage. Various histone modifications, mainly acetylation and methylation, influence DNA exposure to the potential damaging agents and various rearrangements (Zhu and Wani 2010).

The chromatin condensation level is regulated at DNA and histone levels, and changes in DNA methylation and histone modifications are often interdependent. Effector proteins such as the HETEROCHROMATIN PROTEIN1 (HP1) are recruited to the methylated histones: HP1 binds the methylated H3K9 and spreads heterochromatin to the adjacent chromosomal regions (Eskeland et al. 2007). This interaction is especially important for a response to stress. For example, the *Arabidopsis* homologue of HP1, the HETEROCHROMATIN PROTEIN1 (LHP1), is involved in regulating the flowering time in response to environmental cues.

DNA methylation may also serve as the initiation of chromatin condensation marks, since various chromatin modifiers are recruited to the methylated cytosines, as for example, in the case of methyl-CpG-binding domain proteins (MBDs) (Reyes et al. 2002) or the HP1 protein (Jin et al. 2011). The binding of the methylated cytosines by MBDs or HP1 recruits enzymes that modify core histone proteins and change the local chromatin structure.

3.1.1 Heterochromatin Decondensation in Response to Stress

The maintenance of the heterochromatin structure via chromatin condensation is critical for transcriptional gene silencing of repetitive elements. The removal of nucleosomes from specific genomic locations in response to stress could be both an active and a passive process. The original nucleosome positioning and modifications of histones are restored reasonably quickly upon the recovery from stress, suggesting that the removal of nucleosomes and their reloading could indeed be an active process. However, a sequence-specific loss of nucleosomes can also be associated with replication and transcription processes and therefore represent a passive process. It was shown that a long-term exposure to heat of *Arabidopsis thaliana* plants activates repetitive elements, and such activation occurred without DNA methylation loss and with only minute changes in modifications of histones (Pecinka et al. 2010). The transcriptional activity of repetitive elements was associated with the loss of nucleosomes and the decondensation of heterochromatin. When plants recovered from stress, nucleosomes at the regions were restored, and transcriptional silencing was reestablished. The role of chromatin remodeling factors in response and recovery to stress was demonstrated by the fact that chromatin assembly factor 1 (CAF-1) mutants impaired in chromatin assembly functions had a considerably delayed recovery stage and was associated with the nucleosome loading activity (Pecinka et al. 2010).

Curiously, the heterochromatin dissociation was maintained even when silencing and nucleosomes had been reinstalled after the recovery phase; the loss of heterochromatin lasted in the exposed leaves until they started to show signs of senescence. A similar heterochromatin decondensation was observed in *Arabidopsis* plantlets in response to cell culturing. The loss of heterochromatin also occurred in older plants upon floral transition in development, although the heterochromatin decondensation did not activate transcription at the repetitive loci. Thus, sequence-specific heterochromatization and the loss of heterochromatin occur during normal physiological and developmental processes and in responses to stress. Indeed, when plants were exposed to low-light stress, the heterochromatin decondensation was more permanent and was directed toward areas with repetitive elements (Tessadori et al. 2009).

The heterochromatin decondensation of genomic repeats may be a common response to stress, regardless of its nature. The work of Pecinka et al. (2010) puts doubts on this; no heterochromatin decondensation was observed in response to freezing (-4 °C for 24 h) or UV-C irradiation (3000 J/m^2). Also, the heterochromatin decondensation in response to heat stress seems not to occur equally in all tissues;

for example, the nuclei of meristematic cells do not undergo the heat-induced decondensation. Assuming that heat stress is transient in nature—the response to heat stress at the chromatin level should also be transient in nature and should mainly occur in somatic tissues; the absence of changes in the meristem suggests that plants attempt to minimize genetic and epigenetic changes in the germ line and in the progeny. This finding further supports the hypothesis that heterochromatin decondensation is a regulated process that occurs in a specific development stage or in response to specific stresses, and even in this case, it may occur in one type of cells (somatic) and not in others (gametic).

3.1.2 Changes in the Activity of Chromatin Remodeling Factors in Response to Stress

Genome stability in plants is in part regulated through the activity of chromatin-remodeling factors interacting with methylated DNA methylation or modified histones. A member of the SWI2/SNF2 family, the DNA DECREASED DNA METHYLATION1 (DDM1) protein, is a helicase involved in the control of DNA repair, recombination, gene expression and replication (Havas et al. 2001). In *Arabidopsis*, DDM1 regulates DNA methylation via the interaction with MBDs and changes in histone methylation; *ddm1* plants show a disrupted localization of MBDs at chromocenters, suggesting that DDM1 may facilitate the localization of MBDs in specific nuclear domains (Zemach et al. 2005). The role of DDM1 in the control of DNA methylation is shown by the fact that the *ddm1* mutant exhibits an approximately 70% reduction in global genome DNA methylation (Jeddeloh et al. 1999). Demethylation activates transposons and retrotransposons and various silent disease-resistance gene arrays as well as the profound phenotypic instability that worsens when *ddm1* plants are self-propagated. The phenotypic instability in self-propagated *ddm1* plants may be due to the inheritance of induced hypomethylation through mitotic and meiotic cell divisions (Kakutani et al. 1999). RNA-direct DNA methylation (RdDM) is likely one of the mechanisms of the maintenance of DNA methylation via DDM1 when the triggering RNA signal is removed. Although the data on genome instability in *ddm1* is scarce, one can hypothesize that the genome of *ddm1* is unstable since plants have the increased activity of transposons and retrotransposons. *ddm1* plants are more sensitive to a variety of stresses and appear to have a higher frequency of double-strand breaks (Yao et al. 2012a).

The MAINTENANCE OF METHYLATION 1 (MOM1) protein is another chromatin modifier involved in stress response that likely contributes to genome stabilization. MOM1 is involved in silencing of repetitive sequences in *Arabidopsis* by preventing the transcription of 180-bp satellite repeats of transposons (Vaillant et al. 2006). A release of silencing in *mom1* occurs without reducing/alternating the levels of DNA methylation and histone methylation. Although MOM1 is involved in chromatin remodeling, the mutant is not hypersensitive to the DNA-damaging agent methyl methanesulfonate (MMS).

BRU1 is yet another nuclear protein involved in the maintenance of chromatin structure; *bru1* plants are highly sensitive to DNA-damaging agents and exhibit a higher HR frequency (Takeda et al. 2004). The MIM1 protein involved in the maintenance of chromosome structure may also be required for the efficient HR; *mim1* plants have a significantly increased sensitivity to DNA-damaging agents.

DRD1 is another plant-specific SWI/SNF-like protein required for RNA-directed de novo DNA methylation (Kanno et al. 2004). DRD1 interacts with subunits of a novel plant-specific RNA polymerase, pol IVb, NRPD1b, and NRPD2a. Together, DRD1 and the pol IVb complex act downstream of the small RNA (smRNA) biogenesis pathway. DRD1 has been shown to be necessary for the loss of de novo DNA methylation at the target loci after the RNA silencing trigger is withdrawn. Indeed, among DRD1 targets, there are two DNA glycosylases, ROS1 and DME, both of which are involved in active DNA demethylation. The downregulation of *ROS1* in *drd1* and *pol IVb* mutants confirms the importance of the DRD1/pol IVb pathway for an active loss of induced de novo DNA methylation (Penterman et al. 2007). Thus, DRD1 directs reversible silencing of euchromatic promoters in response to RNA signals possibly through the recruitment of DNA methyltransferases for methylation of targeted DNA sequences.

3.2 The Maintenance of Plant Genome Stability Through Changes in DNA Methylation

DNA methylation is the best-described mechanism of regulation of the chromatin structure and gene expression. Stress may result in both hypo- and hypermethylation of the genome, and these changes may represent either a short-term change or a long-term strategy in response to stress (Uthup et al. 2011). For example, promoters of stress-responsive genes may be hypomethylated (Wada et al. 2004; Yao and Kovalchuk 2011), whereas methylation at other genomic loci may not be altered.

Changes in DNA methylation in response to stress may occur due to many different mechanisms, including the activity of DNA methyltransferases; DNA demethylases such as ROS1, DME1, DML2, and DML3; a passive loss of methylation via the exclusion of DNA methyltransferases from the nucleus; changes in the activity of chromatin remodeling factors and effector proteins; and many other changes in proteins regulating the chromatin structure. The regulation of methylation is a complex process, and the absence of one or several DNA methyltransferases does not necessarily result in a total loss of DNA methylation. In the *met1* plants that lack the maintenance of the methyltransferase activity, global hypomethylation is accompanied by hypermethylation at multiple transposon and repetitive element loci. The expression of DNA demethylases, *DME* and *ROS1,* is repressed in the *met1* mutant, likely as an overcompensation mechanism that prevents more extensive losses in methylation. This leads to an increase in de novo non-CG methylation at non-repetitive loci and an increase in the level of RdDM-directed hypermethylation of DNA repeats (Saze et al. 2008).

Methylation is one of the most versatile epigenetic mechanisms of stress response in plants. The immediate stress response of plant somatic tissues results in changes in methylation of various areas of the genome, with genes involved in stress response being primarily hypomethylated. Exposure to cold causes demethylation and transcriptional activation of the *ZmMI1* gene in maize seedlings; the *ZmMI1* gene contains a retrotransposon-like sequence, and its activation mirrors cold-induced root-specific demethylation in the *Ac/Ds* transposon regions followed by their activation (Steward et al. 2000). Hypomethylation in tobacco plants with *NtMET1* antisense results in the upregulation of over 30 genes related to stress response (Wada et al. 2004). Also, the pathogen-responsive gene *NtAlix1* undergoes demethylation and activation in response to viral infection.

The relationship between gene expression and DNA methylation was shown in hypomethylated transgenic tobacco plants expressing an anti-DNA methyltransferase sequence (Choi and Sano 2007). One of the identified genes encoding the glycerophosphodiesterase-like protein (NtGPDL) was earlier reported to be responsive to aluminum stress. Indeed, when the detached leaves from wild-type tobacco plants were treated with aluminum, *NtGPDL* transcripts were demethylated at CCGG sites within 1 h and were induced within 6 h. A similar demethylation pattern was observed in response to salt and low-temperature stress but not to pathogens, indicating a certain stress specificity of response (Choi and Sano 2007).

Changes in the DNA methylation pattern were also observed in other plant-stress interactions. A twofold increase in the level of CpNpG methylation in response to high salinity was found in the nuclear genome of *M. crystallinum* plants (Dyachenko et al. 2006). A correlation between an age-dependent increase in methylation and resistance to the blight pathogen *X. oryzae* in rice was also proposed (Sha et al. 2005). Viral infection of tomato plants resulted in changes in DNA methylation in the genomic regions associated with stress response and stress defense. In hemp and clover, hypomethylation of several marker loci was found in response to heavy metal stress (Panella et al. 2004). Exposure of dandelions to salicylic acid resulted in genome-wide and possibly stress-specific changes in DNA methylation in the exposed plants; these changes were also transmitted to the progeny (Verhoeven et al. 2010).

3.2.1 The Correlation Between DNA Methylation Levels and Genome Stability

DNA methylation is typically associated with a more condensed chromatin, and therefore regions with higher methylation are likely more protected from DNA damage and have a lower mutation frequency. Is this actually true?

Few research papers indicate an inverse correlation between DNA methylation levels and DNA repair activity. DNA methylation suppresses the occurrence of homologous recombination between dispersed sequences, restricting recombination events to the gene-rich regions with a lower level of methylation (Maloisel and Rossignol 1998; Khrustaleva et al. 2005). Also, in *Hevea brasiliensis*, an inverted correlation between DNA methylation levels and DNA rearrangements was observed (Uthup et al. 2011). At the same time, in a study, no significant correlation between the level of methylation and homologous recombination frequency in the Arabidopsis genome has been found (Mirouze et al. 2012). Moreover, the progeny of crosses between wild-type and *met1* mutant *Arabidopsis* plants impaired in the maintenance of CpG methylation exhibited the increased meiotic recombination frequency in the hypomethylated chromosome arms but not in the hypomethylated heterochromatic pericentromeric regions (Mirouze et al. 2012).

There is still no clear inverse correlation established between DNA methylation and genome stability for every studied locus. A certain degree of correlation between methylation of specific cytosine nucleotides and the frequency of point mutations at these sites can however be established. Methylated cytosines are prone to spontaneous deaminations upon which they are converted into thymines, leading to C to T transition mutation. Not surprisingly, CG pairs occur less frequently as compared to other pairs of nucleotides. The analysis of the rate of mutations in *Arabidopsis* plants self-propagated for 30 generations showed that many mutations were C to T transitions (Ossowski et al. 2010). Such bias can only be explained by a high frequency of spontaneous deamination of methylated cytosines. In contrast, the deamination of non-methylated cytosines results in the formation of uracils that are removed by the DNA repair machinery. DNA methylation can also protect DNA from incision and reinsertion of transposons; cytosine methylation prevents the cleavage by many endonucleases, thus preventing the self-incision of transposons. Thus, a higher level of DNA methylation at certain loci may function as a defense mechanism against foreign invasive DNA molecules and as a protection against the cell's own transposable elements.

A higher frequency of deletions/insertions of transposons at long terminal repeats correlates with DNA hypomethylation in wheat (Kraitshtein et al. 2010). Also, the progeny of tobacco plants infected with tobacco mosaic virus had a higher frequency of rearrangements at the *R*-gene-like loci paralleled by hypomethylation at these loci (Boyko et al. 2007).

An important role of DNA methylation for the regulation of gene expression and genome stability is reflected by the fact that plants evolved a specific enzyme to excise methylated cytosines from DNA. ROS1 is a DNA glycosylase with a high affinity to methylated cytosines; ROS1 excises methylated cytosines through the process of base excision repair (Zhu 2009). This enzyme is unique in plants since it functions as a DNA repair enzyme and as an active DNA demethylator; the methylation levels at CpNpG and CpNpN sites increase, and the expression of many transposable elements decreases in the *ros1* mutant (Lei et al. 2015). Active DNA demethylation allows plants to efficiently respond to the developmental and environmental cues through an active methylation and demethylation process. More detailed information on the activity of demethylation processes can be found in the review by Zhu (2009).

3.2.2 Changes in Transposon Activity in Response to Stress and in Response to Changes in DNA Methylation

The activation of transposons in response to stress is very common and is often associated with DNA demethylation. Temperature stresses are the most common stresses altering the transposon activity. Exposure to cold decreases DNA methylation by increasing the rate of excision of the *Tam3* transposon (Hashida et al. 2003). The *Tam3* transposase binds the GCHCG (H = not G) sequence immediately after DNA replication, thus preventing de novo methylation of this sequence. Various stresses also activate the *Tos17* (rice) (Hirochika et al. 1996), *Tto1* (tobacco) (Takeda et al. 1999), *Tnt1* (tobacco) (Beguiristain et al. 2001) and *BARE-1* (barley) (Kalendar et al. 2000) retrotransposons. Pathogen elicitors induce the excision of subfamilies of *Tnt1* retrotransposons in a tissue-specific manner (Beguiristain et al. 2001). It was hypothesized that the activation of *Tnt1* and *Tto1* retrotransposons is likely due to the fact that their promoters have homology to promoters of genes involved in stress response; stress may trigger the activation of transcription factors that bind stress-responsive genes and at the same time bind transposon promoters; the binding to promoters requires local demethylation and decondensation of heterochromatin. Thus, it can be hypothesized that adaptive processes in plants and plant genome evolution may occur through the simultaneous activation of stress-responsive genes and retrotransposons. Indeed, this hypothesis was supported by the finding that in rice plants exposed to cold, the *mPing* element transposed into a rice homologue of the flowering time gene *CONSTANS* (Jiang et al. 2003), thus leading to the alteration of flowering time in the progeny of stressed plants.

Another group of genes altered by transposable elements is the cluster of resistance genes (*R*-genes). These genes are involved in pathogen recognition and resistance due to a specific gene-for-gene interaction. As pathogens try to avoid the recognition through mutations of avirulence (*Avr*) genes, plants are forced to use the same procedure with *R*-genes. Thus, there is a constant arm race between pathogens and plants. There exist many mechanisms of *R*-gene evolution, including homologous recombination and transposition. It is suggested that a number of transposable elements and their derivatives that are present at the *R*-gene loci play a significant role in a rapid diversification of this gene family (Ronald 1998). It would be curious to know whether *R*-genes enjoy a higher frequency of diversification because of the presence of transposons in their sequences or *R*-genes can be diversified because transposons are nonrandomly integrated into the *R*-gene coding areas. The aforementioned reports support a long-standing hypothesis proposed by Barbara McClintock that all kinds of stresses can potentially reshape plant genomes via transposon activation (McClintock 1984).

3.3 The Maintenance of Genome Stability Through Histone Modifications

Proteins that establish histone modifications or interact with them are classified as writers, readers, and erasers (Fig. 1) (Falkenberg and Johnstone 2014). Writers represent enzymes that directly modify histones leading to histone acetylation,

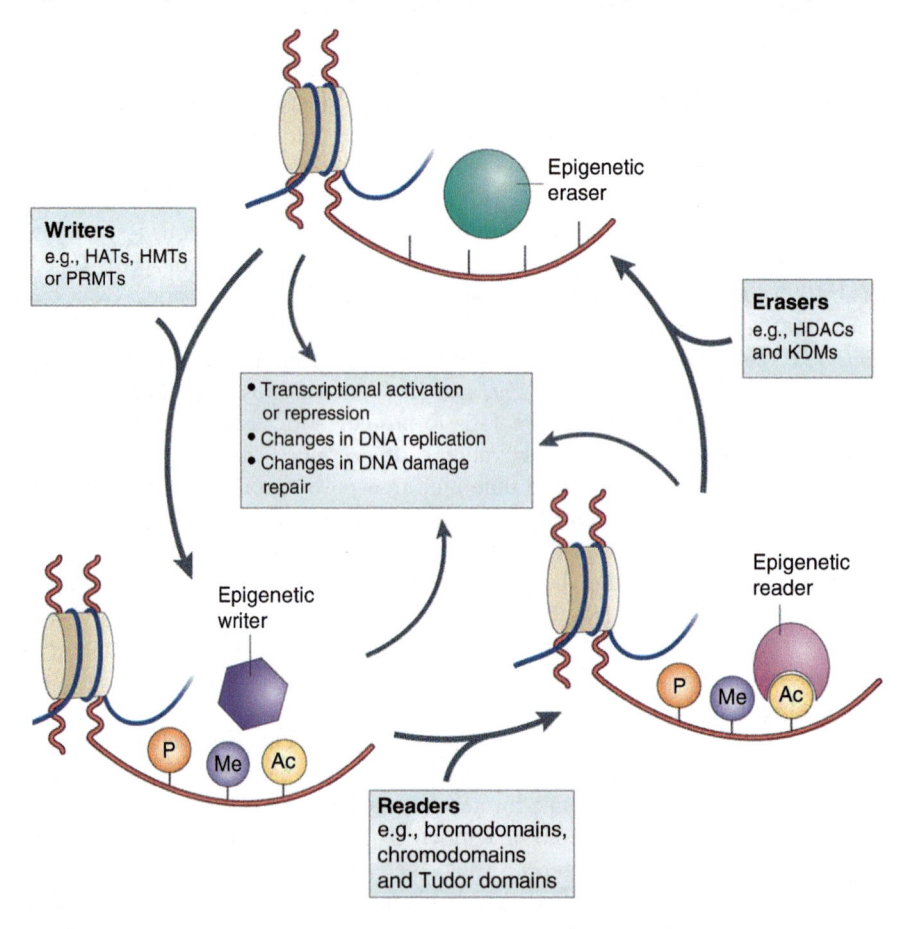

Fig. 1 Epigenetic writers, readers, and erasers. The epigenetic regulation is a dynamic process. Epigenetic writers such as histone acetyltransferases (HATs), histone methyltransferases (HMTs), protein arginine methyltransferases (PRMTs), and kinases lay down epigenetic marks on amino acid residues on histone tails. Epigenetic readers such as proteins containing bromodomains, chromodomains, and Tudor domains bind to these epigenetic marks. Epigenetic erasers such as histone deacetylases (HDACs), lysine demethylases (KDMs), and phosphatases catalyze the removal of epigenetic marks. The addition and removal of these posttranslational modifications of histone tails lead to the addition and/or removal of other marks in a highly complicated histone code. Together, histone modifications regulate various DNA-dependent processes, including transcription, DNA replication, and DNA repair. Reproduced with permission from Falkenberg and Johnstone (2014)

methylation, and phosphorylation; these marks establish local changes in chromatin relaxation or compaction. Readers are proteins containing bromodomain, chromodomain, or Tudor domains involved in recruiting other chromatin modifiers or erasers of the established chromatin marks. Erasers represent proteins such as histone deacetylases (HDACs) and histone demethylases (KDMs) that reverse chromatin marks.

In plants, the transcriptionally active chromatin is represented by an enhancement of H3 and H4 acetylation and trimethylation of lysine 4 from histone H3 (H3K4me3), whereas the silent chromatin is represented by hypoacetylated H3 and H4, methylated lysine 27 (H3K27), and lysine 9 of histone H3 (H3K9) (Kim et al. 2015). Histone acetyltransferases (HATs) and histone deacetylases modulate the expression of genes involved in the regulation of various developmental stages or in response to stress.

Modifications of histones regulate genome stability in many different ways. Since the chromatin structure depends on the association of DNA with specifically modified histones, it is possible that repressive histone marks, such as H3K9me and H3K27me, stabilize the genome, whereas permissive chromatin marks, such as H3K4me, H3K36me, and H3K9ac, may be associated with genome instability. It remains to be shown whether this is the case by a detailed analysis of mutation rates in open or closed chromatin.

A proper response to stress and DNA damage requires high rates of exchange of various histone modifications in order to allow a more efficient access to damaged DNA or a more efficient transcription of loci encoding DNA repair factors. It was demonstrated that loci with higher transcription rates have a higher frequency of production of DSB-induced ncRNAs (diRNAs) from genomic regions with strand breaks (Gao et al. 2014).

Various histone variants are associated with DNA repair mechanisms and genome stability. One of the critical steps in the recognition of a strand break and the assembly of DSB repair factors around a strand break involves phosphorylation of a specific histone variant H2AX resulting in the formation of γH2AX foci (Friesner et al. 2005).

All three types of chromatin modifiers—writers, readers, and erasers—are critical for a proper response to both stress and DNA damage. HDACs, for example, have been implicated in defense against pathogens. HC-toxin from *Cochiobolus carbonum* specifically targets the HDAC activity causing histone hyperacetylation in susceptible corn cultivars (Brosch et al. 1995). HC-toxin inhibits the reduced potassium dependency protein 3/histone deacetylase 1 (RPD3/HDA1) and HD2 classes of proteins (Alvarez et al. 2010). In *Arabidopsis*, the *AtHDAC19* gene has been shown to be induced in a similar manner by the fungus *Alternaria brassicicola* and by exposure to jasmonic acid (Zhou et al. 2005). Silencing of the *AtHDAC19* gene increases plant susceptibility to fungal infection, whereas the overexpression enhances the resistance likely through the activation of the ethylene responsive factor 1 (ERF1). The expression of another HDAC, AtHDAC6, has also been shown to be induced by JA application (Zhou et al. 2005). This enzyme has been shown to affect transgene silencing and DNA methylation. *AtHDAC19* may be broadly required for a response to various pathogens since it has been shown to be required for a response to bacterial pathogen *Pseudomonas syringae*. The proposed mechanism may involve a decrease in histone acetylation through interactions of HDAC19 with WRKY38 and WRKY62, two transcription factors that repress the SA pathway (Kim et al. 2008). The locus-specific suppression of transcription of these two WRKY genes results in the activation of the SA-dependent pathway and thus in the resistance to bacterial pathogens.

3.4 The Regulation of Genome Stability and DNA Repair Through ncRNA Activity

Noncoding RNAs (ncRNAs) are RNA molecules that are not coding for any protein. There is a great variety of ncRNAs with a versatile function in prokaryotes and eukaryotes. Many of them play either a direct or an indirect role in the maintenance of genome stability and the regulation of DNA repair. Noncoding RNAs appear to be very essential for cellular efforts in the maintenance of genome stability in both prokaryotes in which ncRNAs are involved in the regulation of genomic organization and protection against bacteriophages and eukaryotes in which ncRNAs are essential for the regulation of genome size, chromatin structure and compaction, genome integrity, and the efficiency of DNA repair. ncRNAs in the regulation of genome stability may play a direct role by interfering or aiding the process of DNA repair in a direct manner, whereas an indirect role consists in targeting and changing the expression level of genes involved in DNA repair. A direct role of ncRNAs in DNA repair is supported by the fact that several of them are able to interact with DNA repair proteins such as 53BP1 (Pryde et al. 2005), BRCA1 (Ganesan et al. 2002), and Ku70 (Yoo and Dynan 1998). Moreover, RNA-binding proteins have been shown to be recruited to the site of DNA damage and influence the repair efficiency (Adamson et al. 2012; Polo et al. 2012).

Among ncRNAs regulating DNA repair and genome stability are microRNAs (miRNAs) (Maes et al. 2008), small interfering RNAs (siRNAs) (Napoli et al. 1990), Piwi-interacting siRNAs (piRNAs) (Aravin et al. 2001), QDE-2 interacting small RNAs (qiRNAs) (Lee et al. 2009), double-strand break-induced small RNAs (diRNAs) (Wei et al. 2012), and many more. A more detailed role of ncRNAs in the regulation of DNA repair and genome stability is covered in Chapter 25 of the book *Genome Stability* (Kovalchuk 2016).

4 Transgenerational Effects in Response to Stress

4.1 Potential Mechanisms of Transgenerational Effects

Transgenerational response is a phenomenon in which organisms exhibit changes in the progeny in response to the adverse environment experienced by their parents (Boyko and Kovalchuk 2008). Transgenerational effects are exhibited at many levels: DNA methylation and histone modifications, changes in transcriptome, including mRNA and ncRNA transcripts, changes in metabolome and proteome as well as in stress tolerance and genome stability, etc. (reviewed in (Boyko and Kovalchuk 2011; Herman and Sultan 2011; Kinoshita and Seki 2014). Such changes may or may not be heritable, that is, they may or may not be maintained in the next generation when stress is removed. Although it is still unclear what changes are considered to be heritable, we suggest the alterations that persist for two consecutive generations, S1 and S2, where S1 is the first progeny of plants exposed to stress and S2 is the second generation of stressed plants. Nonheritable transgenerational changes are typically those ones that last for a single generation after stress exposure and disappear in the next generation if stressful conditions are not maintained. Most commonly, such changes occur due to differential seed viability/quality caused by the accumulation of metabolites/nutrients that give a certain advantage to plants grown under specific environmental conditions.

The heritability of transgenerational changes is likely dependent on the establishment of certain epigenetic modifications. Similarly to animals, in plants, an early development reprograms many epigenetic modifications such as DNA methylation and histone modifications accumulated during sporophyte development. However, in contrast to animals, reprogramming is less dramatic in plants. In plants, many newly established marks can be retained and passed on to progeny, whereas in animals, the majority (70–90%) of DNA methylation marks are erased (Migicovsky and Kovalchuk 2012). Therefore, epigenetic marks are likely responsible for passing the memory of stress exposure across generations.

For a long time, heritable transgenerational changes, often referred to as soft inheritance, were believed to be extremely rare, whereas hard inheritance (or Mendelian inheritance) that requires mutations to occur in order to introduce a new trait was considered to be the only possible mechanism of inheritance. Such mutations would have to be beneficial to have a chance of becoming fixed in a population. Since mutations are extremely rare, new traits/species emerge rarely and may require many generations to become common in a certain population (Youngson and Whitelaw 2008; Boyko and Kovalchuk 2011; Mirouze and Paszkowski 2011). In contrast, soft inheritance allows an immediate response to the environment, and it is flexible (reversible) allowing the population to respond to the environment frequently and efficiently.

Moreover, it is possible that epigenetic modifications triggered by the environmental stimuli are converted to genetic changes (Fig. 2). For example, cytosine hypomethylation or the establishment of permissive chromatin marks can lead to the increased frequency of genomic rearrangements, whereas cytosine hypermethylation may result in the increased frequency of C to T point mutations due to the frequent deamination of methylated cytosine (Becker et al. 2011; Schmitz et al. 2011). Therefore, the appearance of new traits and new species in response to the changed environmental conditions could be largely driven by epigenetic mechanisms, while genetic mechanisms fix such changes in the genetic code (Fig. 2).

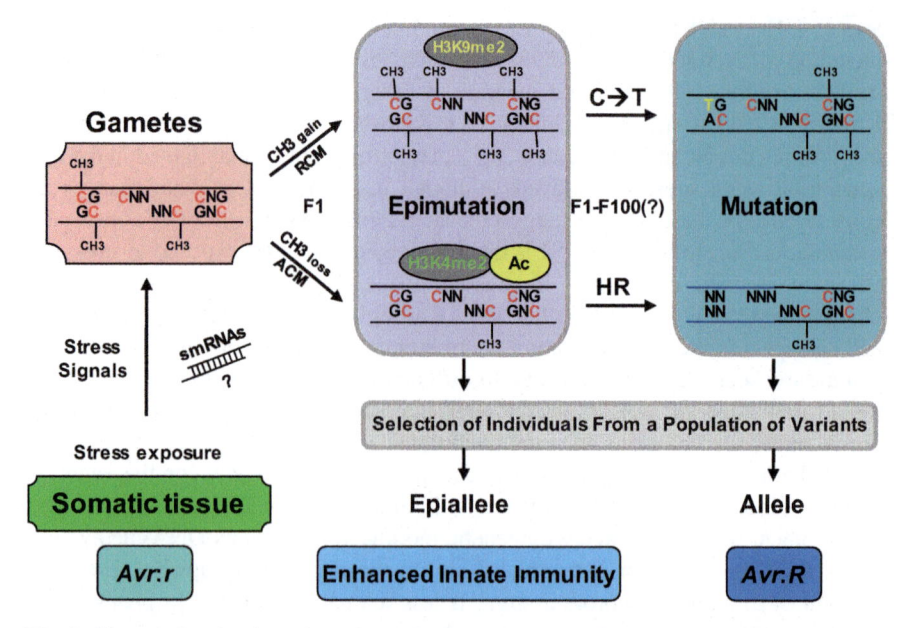

Fig. 2 Stress-induced epigenetic and genetic changes—an evolutionary perspective. In the proposed scenario, stress generates mobile signals, for example, smRNAs, that can reach gametes and influence DNA methylation patterns. The loss or gain of DNA methylation accompanied by repressive chromatin marks (RCM) or active chromatin marks (ACM) represent epimutation events. The diagram shows three types of cytosine methylation, CpG, CpNpG, and CNN. H3K9me2 exemplifies the repressive chromatin mark, whereas H3K4me2 and "Ac" (acetylation) exemplify the active chromatin marks. The hypermethylated regions are prone to a higher frequency of C to T mutations, whereas the hypomethylated regions have a higher frequency of homologous recombination. It is not clear how many generations are required to translate epigenetic mutations into stable genetic ones. Individuals with (epi)mutations that are beneficial for the growth in the specific environment have better chances to survive and reproduce. Thus, new epialleles and alleles are established in the population. The lower panel applies our scenario to plant–pathogen interactions. It can be hypothesized that compatible pathogen interactions in which plants do not have a functional *R*-gene (*Avr:r*) result in a cascade of the abovedescribed events. In the short term, epimutations/epialleles allow plants to withstand pathogen encounters through the enhanced innate immunity. A long-term strategy requiring exposure to the same pathogen over multiple generations leads to the production of new resistance genes (*Avr:R*) as well as the resistance to pathogens due to incompatible interactions

4.2 Transgenerational Response to Abiotic Stress

Abiotic stress is a stress of a non-biological origin, including temperature changes, water availability, wind impacts, and exposures to toxic chemicals or radiation (UV, gamma, etc.), among others. Plants respond to stress at many levels, with the main emphasis on the mechanisms of stress survival and reproduction. Some plants are able to tolerate chronic or recurring abiotic stress in the processes known as adaptation and acclimation (Hasanuzzaman et al. 2013; Tamang and Fukao 2015). These processes operate in somatic cells but may also affect the progeny. Many abiotic stresses destabilize the genome in a direct or an indirect manner (Lebel et al. 1993; Boyko et al. 2006a; Yao and Kovalchuk 2011; Yao et al. 2012b).

Many publications demonstrated that a stress-induced increase in the frequency of somatic HR can be inherited (Boyko et al. 2006a, 2007; Yao and Kovalchuk 2011; Yao et al. 2013a; Bilichak et al. 2012; Boyko et al. 2010). Transgenerational changes in the HR frequency do not occur all the time, and they likely depend on many parameters such as tests/measurements utilized, stress conditions, plant species used, etc. For example, one of the earlier works by Molinier et al. (2006) demonstrated that a single exposure of *Arabidopsis thaliana* plants to stress of UV radiation (UVC, specifically) results in the increased frequency of somatic HR in four consecutive non-stressed generations (Molinier et al. 2006), representing heritable epigenetic inheritance. However in contrast, works by Boyko et al. (2010), Kathiria et al. (2010), and Rahavi et al. (2011) showed that the increased HR frequency is observed mainly in the immediate progeny of stressed plants, and it nearly always drops to a pre-stress level in the next generation when stress is not maintained. It is possible that the persistence of changes in the frequency of HR observed by Molinier et al. (2006) is a unique feature of a particular transgenic line used (Molinier et al. 2006). For example, the transgenic *Arabidopsis thaliana* line, in which an increase was observed, was found to be very unstable without any stress exposure, and a simple propagation of these plants for several generations under normal conditions resulted in a dramatic increase in recombination frequency (Molinier et al. 2006). Also, a work by Pecinka et al. (2009) actually shows that a transgenerational increase in the HR frequency occurs only in specific transgenic *Arabidopsis* lines tested and only in response to a few tested stresses. Specifically, the analysis of recombination frequency in response to ten different stresses showed that transgenerational changes occurred only in response to 2–3 stresses, and these changes were low and stochastic (Pecinka et al. 2009).

One possible explanation for such discrepancy observed in transgenerational changes in the HR frequency could be the intensity of stress used for the analysis. It is possible that only a mild stress may lead to the inheritance of changes in the recombination frequency because a severe stress may have a significant negative effect on plant physiology, somatic cell death, and the negation of epigenetic factors that otherwise would lead to changes in the recombination frequency in progeny. This hypothesis was confirmed by a study that analyzed changes in the HR frequency in response to NaCl; whereas a transgenerational increase in the recombination frequency was most prominent in response to 25 mM NaCl, it was milder in response to 75 mM, and it did not exist in response to 100 mM (Boyko and Kovalchuk 2010). This observation is actually consistent with the results published by Pecinka et al. (2009). The existence of response to mild rather than harsh environmental conditions is reminiscent of the long-known phenomenon of hardening in plants. Hardening in plants describes the increased tolerance to a severe stress when plants experienced a mild stress prior to exposure to a severe stress (Strimbeck et al. 2015). In the case of a transgenerational response, this phenomenon may be referred to as transgenerational hardening.

Our work demonstrated that in most cases, transgenerational changes in the recombination frequency occurred only in the immediate progeny; only two stresses tested (25 mM of salt and UVC) increased the recombination frequency in two consecutive generations, and changes in the second generation were smaller than those in the first one (Boyko et al. 2010).

Would the recombination frequency increase more if plants were propagated in the presence of stresses for more than one generation? Would such changes last longer? The answers to these questions were in part obtained from the work of Rahavi et al. (2011). The authors studied changes in the recombination frequency in response to heavy metal salts such as Ni^{2+}, Cd^{2+}, and Cu^{2+}. They propagated plants on heavy metal salts for up to five generations, and then stress was removed in each generation, starting after generation one. In most cases, an increase in the number of generations exposed to stress resulted in a higher increase in the recombination frequency, although in many cases, this increase reached the plateau already after the first or second generation of exposure. Propagating the progeny of stressed plants under normal conditions resulted in the decreasing recombination frequency, and in those cases where plants were propagated under stress conditions for more generations, a decrease in the HR frequency was less noticeable. In several cases, propagating plants in the presence of stress for 3–4 generations resulted in an interesting phenomenon—the removal of stress did not decrease the recombination frequency even after two generations of growth under normal conditions. These results indicate that stress memory is inherited, and the memory of stress lasts longer when plants are exposed to stress for more than one consecutive generation (Rahavi et al. 2011).

Transgenerational responses may also depend on the timing of stress application. Since plants establish the germline relatively late during development, exposure to stress early during development may allow to pass on the memory of stress application more efficiently (while cells are transitioning to gametes). In contrast, it is likely that stress exposure later during development when gametes are formed may not lead to the efficient generation of stress memory. This is exactly what was observed in the experiment where *Arabidopsis* plants were exposed to heat, cold, and UVC at different time during development: 7, 14, 21, and 28 days postgermination (dpg). The analysis of the HR frequency showed the highest increase in the progeny of plants exposed at 7 dpg. Similarly, the analysis of plant phenotype in the progeny showed that the progeny of plants exposed at 7 dpg had the largest seeds and the largest leaves when grown under normal conditions and when exposed to stress (Rahavi and Kovalchuk 2013).

What type of transgenerational genome instability is the most common? The analysis of genome instability was analyzed using three different transgenic lines, one to measure the point mutation frequency, another—to measure the homologous recombination frequency and yet another one—to measure microsatellite instability (Yao and Kovalchuk 2011). Exposure to various stresses revealed that changes in the HR frequency were the most prominent among three types of genome instability that we tested. Changes in microsatellite instability occurred in response to UVC, heat, and cold but were less prominent than changes in the recombination frequency. Finally, changes in the frequency of point mutations in the progeny were only observed in response to UVC, but not in response to any other stresses (Yao and Kovalchuk 2011). It would be interesting to analyze why changes in the homologous recombination frequency are affected the most in the progeny. Homologous recombination is a mechanism of crossing-over involved in a physical exchange between sister chromatids during meiosis. Such events result in gross chromosomal rearrangements and are likely the most effective in generating novel alleles (Lieberman-Lazarovich and Levy 2011). It can be hypothesized that transgenerational changes in genome stability in response to stress are directed at genome diversification. In this case, homologous recombination should be a mechanism of choice.

4.2.1 Transgenerational Changes in DNA Methylation

In plants, transgenerational changes in DNA methylation in response to stress are a frequent phenomenon. One of the earliest reports in plants showed that exposure to ionizing radiation results in dose-dependent changes in global genome hypermethylation in the progeny; a higher dose of radiation experienced by parental pine tree plants in Chernobyl increased the level of methylation to a higher extent in the progeny (Kovalchuk et al. 2003b). Similarly to the response to ionizing radiation, exposure to salt, flood, heat, cold, or UVC also leads to hypermethylation in the progeny of *Arabidopsis* plants (Boyko et al. 2010). Continuous exposure to stress does not result in a continuous increase in DNA methylation; rather methylation stays at the same elevated level when stress is maintained. Curiously, when the progeny of salt-stressed plants is propagated under normal conditions, they maintain a higher methylation level even though their recombination frequency and stress tolerance decrease to a pre-stress level (Boyko et al. 2010). We can assume that transgenerational changes are triggered by the differential expression of noncoding RNAs that target various genomic loci to establish differential methylation and differential gene expression leading to changes in stress tolerance (Boyko and Kovalchuk 2011).

Changes in global genome methylation do not reflect what is happening at the specific loci. A detailed analysis of DNA methylation at the level of individual loci in the progeny of salt-stressed plants revealed changes in methylation—hypomethylation and hypermethylation—at many loci related to stress response. For example, promoters of *SUVH2*, *SUVH5*, and *SUVH8* genes that were involved in the regulation of the chromatin structure and the promoter of *ROS1*, a gene that helps demethylate DNA, were hypermethylated, whereas promoters of the stress- responsive genes UVH3, ERF1, TUBG1, RAP2.7 and several other genes were hypomethylated (Bilichak et al. 2012). An essential role of DNA methylation for the establishment of transgenerational stress tolerance was demonstrated by the fact that soaking seeds of the progeny of salt-stressed plants in 5-azaC, a chemical compound that modifies cytosines by preventing methylation, do not allow plants to tolerate a higher level of MMS chemicals and eliminates hypermethylation (Boyko et al. 2010).

4.3 Transgenerational Response to Biotic Stress

Infection with pathogens typically leads to changes in plant physiology, the loss of biomass, the accumulation of protective metabolites, early flowering, a decreased fertility, and many other changes. One of the first documented results showing genome destabilization in response to pathogen infection was demonstrated that infection with *Peronospora parasitica* or treatment with chemicals such as 2,6-dichloroisonicotinic acid (INA) or benzothiadiazole (BTH) increases the homologous recombination frequency in the stressed *Arabidopsis* plants (Lucht et al. 2002). Later on, it was shown that infection of tobacco plants with tobacco mosaic virus (TMV) also leads to an increase in the somatic recombination frequency (Kovalchuk et al. 2003a). The resistance to TMV in tobacco plants is conferred by the presence of the resistance gene *N* that allows the cytoplasmic recognition of the virus. Big Havana cultivars that contain the *N* gene produce a local hypersensitive response that allows plants to localize the virus as well as a systemic acquired resistance response that allows plants to tolerate future exposures to the same or different pathogens. SR1 cultivars that lack the *N* gene are sensitive to TMV infection. The resistance to TMV is also temperature sensitive; at temperatures exceeding 28 °C, cultivars that contain the *N* gene also become sensitive. Our work showed that only infection of sensitive plants, either SR1 or Big Havana plants grown at temperatures higher than 28 °C, results in the increased somatic recombination frequency. Importantly, an increase was observed in tissues that were not infected with the virus. Moreover, grafting virus-free leaves of infected plants onto naïve tobacco plants also led to an increase in the recombination frequency (Kovalchuk et al. 2003a).

Infection with TMV also leads to transgenerational changes; the progeny of infected plants had more plants with a fully recombined luciferase transgene, which indicated that the meiotic recombination frequency was also increased. In addition, we found that the somatic HR frequency in the progeny of infected plants was also increased. A similar effect was also observed in response to another similar virus, oilseed rape mosaic virus (ORMV) (Yao et al. 2013b).

Why does the recombination frequency increase only in plants sensitive to TMV? It can be hypothesized that an increase in the HR frequency may be a mechanism for increasing the diversity of resistance genes—templates for the generation of a novel resistance gene that may allow tobacco plants to resist TMV. The analysis of the SR1 genome showed that SR1 plants contain up to 50 loci that carry over 60% homology to the *N* gene (Boyko et al. 2007). The rearrangement frequency at these loci in the progeny of infected SR1 plants was increased over eightfold compared to the progeny of control plants. No difference was observed in the neutral loci–actin loci, suggesting that the increased rearrangement frequency is locus-specific (Boyko et al. 2007). It can be hypothesized that rearrangements at certain loci are controlled by methylation; hypermethylation may prevent loci that are irrelevant to the response to TMV because of rearrangements, whereas hypomethylation allows for the recombination and may enable a genetic diversity where it is most needed (Fig. 2). Indeed, the R-gene-like resistance loci that exhibited a higher rearrangement frequency were hypomethylated, whereas actin loci were similarly methylated in the progeny of infected plants (Boyko et al. 2007).

Another important change observed in the progeny of infected plants was a higher tolerance to TMV infection as well as a higher tolerance to *Pseudomonas syringae* and *Phytophthora nicotianae*. Thus, the progeny of TMV-infected plants have a certain degree of cross-tolerance to bacterial and fungal pathogens. The ability to delay the viral progression is likely triggered by many factors, but our research showed that these plants had a higher endogenous expression of the *PR1* gene and a higher level of callose deposition (Kathiria et al. 2010).

Several studies confirmed our findings in tobacco and *Arabidopsis*. Luna et al. (2012) showed that the progeny of plants that were repeatedly infected with *Pseudomonas syringae* exhibited a higher tolerance to this pathogen and to fungal pathogen *Hyaloperonospora arabidopsidis* (Luna et al. 2012). A higher pathogen tolerance was also observed in the next generation even when plants were propagated under normal conditions. Also, a single inoculation of *Arabidopsis* with *Pseudomonas syringae* resulted in a stronger and quicker response to *Pseudomonas* infection in the progeny (Slaughter et al. 2012).

Insects are other pathogens that can trigger a transgenerational response. Exposure to herbivores in wild radish plants results in a higher tolerance to herbivory in the progeny (Agrawal 2001). Also, yellow monkeyflower plants exposed to herbivory also generate plants with herbivory tolerance—plants with an increased trichome density (Colicchio et al. 2015; Holeski et al. 2010). The analysis of the progeny of *Arabidopsis* and tomato plants that were exposed to caterpillars showed an enhanced resistance to two out of three herbivores tested (Rasmann et al. 2012). A higher tolerance to herbivores was also observed in the second generation when plants were propagated under normal conditions, but no such tolerance was observed in the third generation (Rasmann et al. 2012).

5 Mechanisms of Transgenerational Inheritance of Stress Memory

What controls a transgenerational response in the form of changes in genome stability? Several mechanisms may be involved, and our experiments indicate a possible role of DNA repair proteins and epigenetic regulators such as ncRNAs and DNA cytosine methylation.

5.1 The Role of DNA Repair Factors

Our previous analysis showed the higher transcript levels of several DNA repair genes in the progeny of stressed plants (Bilichak et al. 2012; Boyko et al. 2010). It is possible that the differential expression of repair genes in plants exposed to stress may influence the recombination frequency in the progeny. It is not clear, however, whether the presence of all repair factors is essential to observe transgenerational changes in genome rearrangements. For example, the recombination frequency is increased in wild-type plants and *atm* and *rad51b* mutants but not in *ku80* plants in response to various abiotic stressors (Yao et al. 2013a). In the progeny, the increased recombination frequency is observed in wild-type and *ku80* plants but not in the *atm* mutant or *rad51b* mutant plants. It is possible that the functional ATM proteins that recognize DSBs and AtRAD51B involved in the homologous recombination pathway to repair such breaks are needed for the initiation of a transgenerational signal or its transmission through gametes. Other proteins, such as KU80 and UVH3, appeared to be dispensable for transgenerational changes (Yao et al. 2013a). It is curious to draw a parallel with recent reports describing the production of diRNAs (reviewed in (Kovalchuk 2016). It was shown that strand breaks trigger the production of diRNA originating from the site of a strand break and that these diRNAs are depleted in the *atm* and *atr* mutants as well as in various mutants impaired in the epigenetic regulation, namely, DCL3, AGO2, and several others (Gao et al. 2014; Wei et al. 2012). Moreover, it was found that the NHEJ process (in which KU80 is known to participate) was not impaired when diRNAs were depleted (Gao et al. 2014; Wei et al. 2012).

5.2 The Role of Epigenetic Factors

The aforementioned reports clearly demonstrate that epigenetic regulators are involved in transgenerational changes in genome instability. The locus-specific changes in DNA methylation in the progeny of stressed plants are likely directed to control the expression of loci that are somehow involved in a transgenerational response.

In plants, DNA methylation occurs at symmetrical cytosines, including CG and CNG sites and asymmetrical cytosines at CNN sites. Whereas symmetrical methylation can be maintained after replication and resynthesis steps, asymmetrical cytosines cannot be maintained and are established through de novo methylation events; such events require the function of ncRNAs through RNA-dependent DNA methylation that functions in a sequence-specific manner. *Arabidopsis* has four Dicer-like (DCL) proteins; only one of them, DCL1, is involved in miRNA biogenesis, whereas the other three, DCL2, DCL3, and DCL4, are involved in biogenesis of small interfering RNAs (siRNAs) (Khraiwesh et al. 2012). RdDM involves the activity of PolIV, DCL3, RDR6, DRM2, and several other proteins together with siRNAs that function in a sequence-specific manner. There are two major RdDM pathways, PolIV-RdDM and RDR6-RdDM (Bond and Baulcombe 2015). We hypothesized that the RdDM pathways may be the pathways required for the establishment of transgenerational changes in methylation and genome stability.

Indeed, we showed that *dcl2* and *dcl3* mutants were partially impaired in a transgenerational increase in the HR frequency and DNA methylation in response to flood, heat, cold, and UVC (Boyko et al. 2010). It was also shown that the *Arabidopsis dcl2*, *dcl3*, and *dcl4* triple mutant did not inherit the resistance to insects in response to parental herbivory, whereas wild-type plants did inherit it (Rasmann et al. 2012). These two reports support the essential role of DCLs and siRNAs in transgenerational changes. In contrast, Ito et al. (2011) showed that DCL3 may not be necessary for the transgenerational transposition of *ONSEN* in response to heat (Ito et al. 2011). The authors hypothesized that DCL3 may be partially restricting the accumulation of *ONSEN* in response to heat stress in somatic tissues but may not be required for such response in the progeny.

We propose several steps of the establishment of a transgenerational response. First, stress response in somatic cells involves the differential expression of mRNAs, ncRNAs, changes in DNA methylation, and histone modifications in somatic tissues. In those cases when stress occurs early during development, it may result in changes in the meristem and thus can be fixed in gametes. If stress occurs when gametes are established, they may also be altered in response to stress. Even if meristem cells or gametes are not altered directly, these cells may acquire the information about stress from somatic cells via the circulating molecules (ncRNAs?) in plasmodesmata and phloem. One possibility is that changes in DNA methylation and histone modifications caused by the RdDM mechanism may occur in the exposed meristem cells. Another possibility is that changes in DNA methylation occur in mature gametes or early embryos and are caused by the differential expression of ncRNAs produced in gametes or embryos or even in the endosperm. Second, changes that occur in meristem cells or in the developing gametes need to escape reprogramming, a mechanism that erases the epigenetic marks such as changes in DNA methylation, histone modifications, and the degradation of mRNAs in pollen. The epigenetic changes caused by stress also need to survive the second level of reprogramming that occurs after the fertilization event. Third, it is possible that transgenerational changes occur directly in the developing progeny triggered by the differentially expressed ncRNAs that survived all reprogramming steps. Our work in *Brassica rapa* showed that heat stress induces changes in ncRNA and mRNA expression in meristem tissues and gametes; some of these changes were propagated into the developing embryo and even into the progeny (Bilichak et al. 2015).

Changes in the HR frequency are likely triggered by changes in DNA methylation or histone marks, but they may also occur due to the differential expression of DNA repair genes (see above). Changes in tolerance to stress could be due to many factors, including the differential accumulation of metabolites, changes in the chromatin structure (due to DNA or histone methylation), changes in the expression of stress-responsive genes, and many other factors.

Fourth, the propagation and maintenance of stress memory to the next generations may require a continuous stress exposure (generation after generation).This is not surprising because if changes in DNA methylation and ncRNAs that trigger them play an essential role, they need to be generated constantly to reinforce a transgenerational memory and replenish molecules depleted during reprogramming. Future research will show whether this theory has any merit.

6 Concluding Remarks

In this chapter, we have summarized various mechanisms regulating genome stability, including the choice of the DNA repair pathway and the role of epigenetic factors such as DNA methylation, histone modifications, and noncoding RNAs. We have described several experimental evidences indicating that the progeny of stressed plants exhibit a variety of changes, including changes in stress tolerance, DNA methylation, and genome stability. We have also demonstrated that transgenerational changes are epigenetically regulated. We hypothesized that transgenerational changes are caused by the differential expression of ncRNAs and RdDM mechanisms causing differential changes in DNA methylation and histone modifications. Direct links between the levels of DNA rearrangements at specific loci and the expression of siRNAs (or other ncRNAs) mapping to these genomic regions remain to be established. It remains to be shown whether such siRNAs are passed on from the progeny via gametes, or their expression is induced in the early stage of embryo development or even in the germinated plants. It is also not very clear whether the chromatin structure has a direct effect on the level of genome rearrangements because such links are not evidently established. Finally, it remains to be shown whether siRNAs that trigger rearrangements at certain genomic loci are actually stress-specific and whether they can direct changes (methylation or DNA rearrangements) at the loci that may somehow assist plants in stress survival.

Glossary

Epigenetic changes Heritable changes in gene expression that do not involve changes in the DNA sequence. Epigenetic changes are typically associated with reversible modifications of DNA (cytosine methylation) or histones (methylation, acetylation, etc.).

Transgenerational responses (effects) Changes in a phenotype (associated with epigenetic or physiological changes) that are apparent in the progeny of plants exposed to adverse conditions.

Nonheritable transgenerational effects (responses) Transgenerational effects that do not persist beyond the immediate generation after stress; the removal of a stressor prevents these changes from being passed on to the next generation. Such effects are typically caused by changes in seed quality due to the accumulation of metabolites, nutrients, etc., giving an advantage to the growing seedling under certain environmental conditions.

Heritable transgenerational effects Transgenerational effects persisting for more than one generation, even when the stimulus causing these changes is removed; these effects are typically associated with changes in the epigenome.

Memory Genetic, epigenetic, or physiological changes that outlast stressful conditions and modify the response to the subsequent stress treatments in the same or the next generation (a transgenerational memory).

Hard or Mendelian inheritance The inheritance of traits via changes in the DNA sequence; according to Mendelian inheritance, new traits can only appear as a result of a mutation.

Soft inheritance The inheritance of traits that does not include mutations and changes in the DNA sequence but rather involves changes in the epigenetic regulation in the form of DNA methylation, histone modifications, etc.

Repressive chromatin marks Posttranslational histone modifications associated with low levels of gene expression and condensed chromatin.

Permissive (active) chromatin marks Posttranslational histone modifications associated with high levels of gene expression and open chromatin.

Hardening An increased tolerance to severe stress in plants after exposure to mild stress, also referred to as priming, acclimation, conditioning, etc.

Transgenerational hardening A higher stress tolerance in the progeny of plants exposed to mild stress; it likely occurs due to the accumulation of nutrients and metabolites in seeds as well as due to changes in the epigenetic regulation.

References

Adamson B, Smogorzewska A, Sigoillot FD, King RW, Elledge SJ (2012) A genome-wide homologous recombination screen identifies the RNA-binding protein RBMX as a component of the DNA-damage response. Nat Cell Biol 14:318–328. https://doi.org/10.1038/ncb2426

Agrawal AA (2001) Transgenerational consequences of plant responses to herbivory: an adaptive maternal effect? Am Nat 157:555–569. https://doi.org/10.1086/319932

Alvarez ME, Nota F, Cambiagno DA (2010) Epigenetic control of plant immunity. Mol Plant Pathol 11:563–576. https://doi.org/10.1111/j.1364-3703.2010.00621.x

Aravin AA, Naumova NM, Tulin AV, Vagin VV, Rozovsky YM, Gvozdev VA (2001) Double-stranded RNA-mediated silencing of genomic tandem repeats and transposable elements in the D. melanogaster germline. Curr Biol 11:1017–1027

Becker C, Hagmann J, Muller J, Koenig D, Stegle O, Borgwardt K, Weigel D (2011) Spontaneous epigenetic variation in the *Arabidopsis thaliana* methylome. Nature 480:245–249. https://doi.org/10.1038/nature10555

Beguiristain T, Grandbastien MA, Puigdomenech P, Casacuberta JM (2001) Three Tnt1 subfamilies show different stress-associated patterns of expression in tobacco. Consequences for retrotransposon control and evolution in plants. Plant Physiol 127:212–221

Bilichak A, Ilnystkyy Y, Hollunder J, Kovalchuk I (2012) The progeny of *Arabidopsis thaliana* plants exposed to salt exhibit changes in DNA methylation, histone modifications and gene expression. PLoS One 7:e30515. https://doi.org/10.1371/journal.pone.0030515

Bilichak A, Ilnytskyy Y, Woycicki R, Kepeshchuk N, Fogen D, Kovalchuk I (2015) The elucidation of stress memory inheritance in *Brassica rapa* plants. Front Plant Sci 6:5. https://doi.org/10.3389/fpls.2015.00005

Bond DM, Baulcombe DC (2015) Epigenetic transitions leading to heritable, RNA-mediated de novo silencing in *Arabidopsis thaliana*. Proc Natl Acad Sci U S A 112:917–922. https://doi.org/10.1073/pnas.1413053112

Boyko A, Kovalchuk I (2008) Epigenetic control of plant stress response. Environ Mol Mutagen 49:61–72. https://doi.org/10.1002/em.20347

Boyko A, Kovalchuk I (2010) Transgenerational response to stress in Arabidopsis thaliana. Plant Signal Behav 5:995–998. https://doi.org/10.1371/journal.pone.0009514

Boyko A, Kovalchuk I (2011) Genome instability and epigenetic modification--heritable responses to environmental stress? Curr Opin Plant Biol 14:260–266. https://doi.org/10.1016/j.pbi.2011.03.003

Boyko A, Hudson D, Bhomkar P, Kathiria P, Kovalchuk I (2006a) Increase of homologous recombination frequency in vascular tissue of Arabidopsis plants exposed to salt stress. Plant Cell Physiol 47:736–742. https://doi.org/10.1093/pcp/pcj045

Boyko A, Zemp F, Filkowski J, Kovalchuk I (2006b) Double-strand break repair in plants is developmentally regulated. Plant Physiol 141:488–497. https://doi.org/10.1104/pp.105.074658

Boyko A, Kathiria P, Zemp FJ, Yao Y, Pogribny I, Kovalchuk I (2007) Transgenerational changes in the genome stability and methylation in pathogen-infected plants: (virus-induced plant genome instability). Nucleic Acids Res 35:1714–1725. https://doi.org/10.1093/nar/gkm029

Boyko A, Blevins T, Yao Y, Golubov A, Bilichak A, Ilnytskyy Y, Hollunder J, Meins F Jr, Kovalchuk I (2010) Transgenerational adaptation of Arabidopsis to stress requires DNA methylation and the function of Dicer-like proteins. PLoS One 5:e9514. https://doi.org/10.1371/journal.pone.0009514

Brosch G, Ransom R, Lechner T, Walton JD, Loidl P (1995) Inhibition of maize histone deacetylases by HC toxin, the host-selective toxin of *Cochliobolus carbonum*. Plant Cell 7:1941–1950. https://doi.org/10.1105/tpc.7.11.1941

Chinnusamy V, Zhu JK (2009) Epigenetic regulation of stress responses in plants. Curr Opin Plant Biol 12:133–139. https://doi.org/10.1016/j.pbi.2008.12.006

Chiruvella KK, Liang Z, Wilson TE (2013) Repair of double-strand breaks by end joining. Cold Spring Harb Perspect Biol 5:a012757. https://doi.org/10.1101/cshperspect.a012757

Choi CS, Sano H (2007) Abiotic-stress induces demethylation and transcriptional activation of a gene encoding a glycerophosphodiesterase-like protein in tobacco plants. Mol Genet Genomics 277:589–600. https://doi.org/10.1007/s00438-007-0209-1

Colicchio JM, Miura F, Kelly JK, Ito T, Hileman LC (2015) DNA methylation and gene expression in Mimulus guttatus. BMC Genomics 16:507. https://doi.org/10.1186/s12864-015-1668-0

Dassler A, Roscher C, Temperton VM, Schumacher J, Schulze ED (2008) Adaptive survival mechanisms and growth limitations of small-stature herb species across a plant diversity gradient. Plant Biol 10:573–587. https://doi.org/10.1111/j.1438-8677.2008.00073.x

Downey M, Durocher D (2006) Chromatin and DNA repair: the benefits of relaxation. Nat Cell Biol 8:9–10. https://doi.org/10.1038/ncb0106-9

Dyachenko OV, Zakharchenko NS, Shevchuk TV, Bohnert HJ, Cushman JC, Buryanov YI (2006) Effect of hypermethylation of CCWGG sequences in DNA of *Mesembryanthemum crystallinum* plants on their adaptation to salt stress. Biochemistry (Mosc) 71:461–465

Eskeland R, Eberharter A, Imhof A (2007) HP1 binding to chromatin methylated at H3K9 is enhanced by auxiliary factors. Mol Cell Biol 27:453–465. https://doi.org/10.1128/MCB.01576-06

Falkenberg KJ, Johnstone RW (2014) Histone deacetylases and their inhibitors in cancer, neurological diseases and immune disorders. Nat Rev Drug Discov 13:673–691. https://doi.org/10.1038/nrd4360

Friesner JD, Liu B, Culligan K, Britt AB (2005) Ionizing radiation-dependent gamma-H2AX focus formation requires ataxia telangiectasia mutated and ataxia telangiectasia mutated and Rad3-related. Mol Biol Cell 16:2566–2576. https://doi.org/10.1091/mbc.E04-10-0890

Ganesan S, Silver DP, Greenberg RA, Avni D, Drapkin R, Miron A, Mok SC, Randrianarison V, Brodie S, Salstrom J, Rasmussen TP, Klimke A, Marrese C, Marahrens Y, Deng CX, Feunteun J, Livingston DM (2002) BRCA1 supports XIST RNA concentration on the inactive X chromosome. Cell 111:393–405

Gao M, Wei W, Li MM, Wu YS, Ba Z, Jin KX, Li MM, Liao YQ, Adhikari S, Chong Z, Zhang T, Guo CX, Tang TS, Zhu BT, Xu XZ, Mailand N, Yang YG, Qi Y, Rendtlew Danielsen JM (2014) Ago2 facilitates Rad51 recruitment and DNA double-strand break repair by homologous recombination. Cell Res 24:532–541. https://doi.org/10.1038/cr.2014.36

Hasanuzzaman M, Nahar K, Alam MM, Roychowdhury R, Fujita M (2013) Physiological, biochemical, and molecular mechanisms of heat stress tolerance in plants. Int J Mol Sci 14:9643–9684. https://doi.org/10.3390/ijms14059643

Hashida SN, Kitamura K, Mikami T, Kishima Y (2003) Temperature shift coordinately changes the activity and the methylation state of transposon Tam3 in *Antirrhinum majus*. Plant Physiol 132:1207–1216

Havas K, Whitehouse I, Owen-Hughes T (2001) ATP-dependent chromatin remodeling activities. Cell Mol Life Sci 58:673–682

Herman JJ, Sultan SE (2011) Adaptive transgenerational plasticity in plants: case studies, mechanisms, and implications for natural populations. Front Plant Sci 2:102. https://doi.org/10.3389/fpls.2011.00102

Hirochika H, Sugimoto K, Otsuki Y, Tsugawa H, Kanda M (1996) Retrotransposons of rice involved in mutations induced by tissue culture. Proc Natl Acad Sci U S A 93(15):7783–7788

Holeski LM, Chase-Alone R, Kelly JK (2010) The genetics of phenotypic plasticity in plant defense: trichome production in *Mimulus guttatus*. Am Nat 175:391–400. https://doi.org/10.1086/651300

Ito H, Gaubert H, Bucher E, Mirouze M, Vaillant I, Paszkowski J (2011) An siRNA pathway prevents transgenerational retrotransposition in plants subjected to stress. Nature 472:115–119. https://doi.org/10.1038/nature09861

Jeddeloh JA, Stokes TL, Richards EJ (1999) Maintenance of genomic methylation requires a SWI2/SNF2-like protein. Nat Genet 22:94–97. https://doi.org/10.1038/8803

Jiang N, Bao Z, Zhang X, Hirochika H, Eddy SR, McCouch SR, Wessler SR (2003) An active DNA transposon family in rice. Nature 421:163–167. https://doi.org/10.1038/nature01214

Jin B, Li Y, Robertson KD (2011) DNA methylation: superior or subordinate in the epigenetic hierarchy? Genes Cancer 2:607–617. https://doi.org/10.1177/1947601910393957

Kakutani T, Munakata K, Richards EJ, Hirochika H (1999) Meiotically and mitotically stable inheritance of DNA hypomethylation induced by ddm1 mutation of *Arabidopsis thaliana*. Genetics 151:831–838

Kalendar R, Tanskanen J, Immonen S, Nevo E, Schulman AH (2000) Genome evolution of wild barley (*Hordeum spontaneum*) by BARE-1 retrotransposon dynamics in response to sharp microclimatic divergence. Proc Natl Acad Sci U S A 97:6603–6607. https://doi.org/10.1073/pnas.110587497

Kanno T, Mette MF, Kreil DP, Aufsatz W, Matzke M, Matzke AJ (2004) Involvement of putative SNF2 chromatin remodeling protein DRD1 in RNA-directed DNA methylation. Curr Biol 14:801–805. https://doi.org/10.1016/j.cub.2004.04.037

Kathiria P, Sidler C, Golubov A, Kalischuk M, Kawchuk LM, Kovalchuk I (2010) Tobacco mosaic virus infection results in an increase in recombination frequency and resistance to viral, bacterial, and fungal pathogens in the progeny of infected tobacco plants. Plant Physiol 153:1859–1870. https://doi.org/10.1104/pp.110.157263

Khraiwesh B, Zhu JK, Zhu J (2012) Role of miRNAs and siRNAs in biotic and abiotic stress responses of plants. Biochim Biophys Acta 1819:137–148. https://doi.org/10.1016/j.bbagrm.2011.05.001

Khrustaleva LI, de Melo PE, van Heusden AW, Kik C (2005) The integration of recombination and physical maps in a large-genome monocot using haploid genome analysis in a trihybrid allium population. Genetics 169(3):1673–1685. https://doi.org/10.1534/genetics.104.038687

Kim KC, Lai Z, Fan B, Chen Z (2008) Arabidopsis WRKY38 and WRKY62 transcription factors interact with histone deacetylase 19 in basal defense. Plant Cell 20:2357–2371. https://doi.org/10.1105/tpc.107.055566

Kim JM, Sasaki T, Ueda M, Sako K, Seki M (2015) Chromatin changes in response to drought, salinity, heat, and cold stresses in plants. Front Plant Sci 6:114. https://doi.org/10.3389/fpls.2015.00114

Kinoshita T, Seki M (2014) Epigenetic memory for stress response and adaptation in plants. Plant Cell Physiol 55:1859–1863. https://doi.org/10.1093/pcp/pcu125

Kovalchuk I (2016) Non-coding RNAs in genome integrity. In: Kovalchuk I, Kovalchuk O (eds) Genome stability, Translational epigenetics. Elsevier, New York, pp 425–443

Kovalchuk I, Kovalchuk O, Kalck V, Boyko V, Filkowski J, Heinlein M, Hohn B (2003a) Pathogen-induced systemic plant signal triggers DNA rearrangements. Nature 423:760–762. https://doi.org/10.1038/nature01683

Kovalchuk O, Burke P, Arkhipov A, Kuchma N, James SJ, Kovalchuk I, Pogribny I (2003b) Genome hypermethylation in Pinus silvestris of Chernobyl--a mechanism for radiation adaptation? Mutat Res 529:13–20

Kraitshtein Z, Yaakov B, Khasdan V, Kashkush K (2010) Genetic and epigenetic dynamics of a retrotransposon after allopolyploidization of wheat. Genetics 186:801–812. https://doi.org/10.1534/genetics.110.120790

Langley CH, Montgomery E, Hudson R, Kaplan N, Charlesworth B (1988) On the role of unequal exchange in the containment of transposable element copy number. Genet Res 52:223–235

Lebel EG, Masson J, Bogucki A, Paszkowski J (1993) Stress-induced intrachromosomal recombination in plant somatic cells. Proc Natl Acad Sci U S A 90:422–426

Lee HC, Chang SS, Choudhary S, Aalto AP, Maiti M, Bamford DH, Liu Y (2009) qiRNA is a new type of small interfering RNA induced by DNA damage. Nature 459:274–277. https://doi.org/10.1038/nature08041

Lei M, Zhang H, Julian R, Tang K, Xie S, Zhu JK (2015) Regulatory link between DNA methylation and active demethylation in Arabidopsis. Proc Natl Acad Sci U S A 112:3553–3557. https://doi.org/10.1073/pnas.1502279112

Lieber MR (2010) The mechanism of double-strand DNA break repair by the nonhomologous DNA end-joining pathway. Annu Rev Biochem 79:181–211. https://doi.org/10.1146/annurev.biochem.052308.093131

Lieberman-Lazarovich M, Levy AA (2011) Homologous recombination in plants: an antireview. Methods Mol Biol 701:51–65. https://doi.org/10.1007/978-1-61737-957-4_3

Liu WF, Yu SS, Chen GJ, Li YZ (2006) DNA damage checkpoint, damage repair, and genome stability. Yi Chuan Xue Bao 33:381–390. https://doi.org/10.1016/S0379-4172(06)60064-4

Lombard DB, Chua KF, Mostoslavsky R, Franco S, Gostissa M, Alt FW (2005) DNA repair, genome stability, and aging. Cell 120:497–512. https://doi.org/10.1016/j.cell.2005.01.028

Lucht JM, Mauch-Mani B, Steiner HY, Metraux JP, Ryals J, Hohn B (2002) Pathogen stress increases somatic recombination frequency in Arabidopsis. Nat Genet 30:311–314. https://doi.org/10.1038/ng846

Luna E, Bruce TJ, Roberts MR, Flors V, Ton J (2012) Next-generation systemic acquired resistance. Plant Physiol 158:844–853. https://doi.org/10.1104/pp.111.187468

Madlung A, Comai L (2004) The effect of stress on genome regulation and structure. Ann Bot 94:481–495. https://doi.org/10.1093/aob/mch172

Maes OC, An J, Sarojini H, Wu H, Wang E (2008) Changes in MicroRNA expression patterns in human fibroblasts after low-LET radiation. J Cell Biochem 105:824–834. https://doi.org/10.1002/jcb.21878

Maloisel L, Rossignol JL (1998) Suppression of crossing-over by DNA methylation in Ascobolus. Genes Dev 12:1381–1389

McClintock B (1984) The significance of responses of the genome to challenge. Science 226:792–801

Migicovsky Z, Kovalchuk I (2012) Epigenetic modifications during angiosperm gametogenesis. Front Plant Sci 3:20. https://doi.org/10.3389/fpls.2012.00020

Mirouze M, Paszkowski J (2011) Epigenetic contribution to stress adaptation in plants. Curr Opin Plant Biol 14:267–274. https://doi.org/10.1016/j.pbi.2011.03.004

Mirouze M, Lieberman-Lazarovich M, Aversano R, Bucher E, Nicolet J, Reinders J, Paszkowski J (2012) Loss of DNA methylation affects the recombination landscape in Arabidopsis. Proc Natl Acad Sci U S A 109:5880–5885. https://doi.org/10.1073/pnas.1120841109

Molinier J, Ries G, Zipfel C, Hohn B (2006) Transgeneration memory of stress in plants. Nature 442:1046–1049. https://doi.org/10.1038/nature05022

Napoli C, Lemieux C, Jorgensen R (1990) Introduction of a chimeric chalcone synthase gene into Petunia results in reversible co-suppression of homologous genes in trans. Plant Cell 2:279–289. https://doi.org/10.1105/tpc.2.4.279

Ossowski S, Schneeberger K, Lucas-Lledo JI, Warthmann N, Clark RM, Shaw RG, Weigel D, Lynch M (2010) The rate and molecular spectrum of spontaneous mutations in *Arabidopsis thaliana*. Science 327:92–94. https://doi.org/10.1126/science.1180677

Panella M, Aina R, Renna M, Santagostino A, Palin L (2004) A study of air pollutants and acute asthma exacerbations in urban areas: status report. Environ Pollut 123:399–402. https://doi.org/10.1016/j.envpol.2003.09.003

Pecinka A, Rosa M, Schikora A, Berlinger M, Hirt H, Luschnig C, Mittelsten Scheid O (2009) Transgenerational stress memory is not a general response in Arabidopsis. PLoS One 4:e5202. https://doi.org/10.1371/journal.pone.0005202

Pecinka A, Dinh HQ, Baubec T, Rosa M, Lettner N, Mittelsten Scheid O (2010) Epigenetic regulation of repetitive elements is attenuated by prolonged heat stress in Arabidopsis. Plant Cell 22:3118–3129. https://doi.org/10.1105/tpc.110.078493

Penterman J, Uzawa R, Fischer RL (2007) Genetic interactions between DNA demethylation and methylation in Arabidopsis. Plant Physiol 145:1549–1557. https://doi.org/10.1104/pp.107.107730

Polo SE, Blackford AN, Chapman JR, Baskcomb L, Gravel S, Rusch A, Thomas A, Blundred R, Smith P, Kzhyshkowska J, Dobner T, Taylor AM, Turnell AS, Stewart GS, Grand RJ, Jackson SP (2012) Regulation of DNA-end resection by hnRNPU-like proteins promotes DNA double-strand break signaling and repair. Mol Cell 45:505–516. https://doi.org/10.1016/j.molcel.2011.12.035

Pryde F, Khalili S, Robertson K, Selfridge J, Ritchie AM, Melton DW, Jullien D, Adachi Y (2005) 53BP1 exchanges slowly at the sites of DNA damage and appears to require RNA for its association with chromatin. J Cell Sci 118:2043–2055. https://doi.org/10.1242/jcs.02336

Rahavi SM, Kovalchuk I (2013) Changes in homologous recombination frequency in Arabidopsis thaliana plants exposed to stress depend on time of exposure during development and on duration of stress exposure. Physiol Mol Biol Plants 19:479–488. https://doi.org/10.1007/s12298-013-0197-z

Rahavi MR, Migicovsky Z, Titov V, Kovalchuk I (2011) Transgenerational adaptation to heavy metal salts in Arabidopsis. Front Plant Sci 2:91. https://doi.org/10.3389/fpls.2011.00091

Rasmann S, De Vos M, Casteel CL, Tian D, Halitschke R, Sun JY, Agrawal AA, Felton GW, Jander G (2012) Herbivory in the previous generation primes plants for enhanced insect resistance. Plant Physiol 158:854–863. https://doi.org/10.1104/pp.111.187831

Reyes JC, Hennig L, Gruissem W (2002) Chromatin-remodeling and memory factors. New regulators of plant development. Plant Physiol 130:1090–1101. https://doi.org/10.1104/pp.006791

Ronald PC (1998) Resistance gene evolution. Curr Opin Plant Biol 1:294–298

Saze H, Sasaki T, Kakutani T (2008) Negative regulation of DNA methylation in plants. Epigenetics 3:122–124

Schmitz RJ, Schultz MD, Lewsey MG, O'Malley RC, Urich MA, Libiger O, Schork NJ, Ecker JR (2011) Transgenerational epigenetic instability is a source of novel methylation variants. Science 334:369–373. https://doi.org/10.1126/science.1212959

Sha AH, Lin XH, Huang JB, Zhang DP (2005) Analysis of DNA methylation related to rice adult plant resistance to bacterial blight based on methylation-sensitive AFLP (MSAP) analysis. Mol Genet Genomics 273:484–490. https://doi.org/10.1007/s00438-005-1148-3

Shibata A, Conrad S, Birraux J, Geuting V, Barton O, Ismail A, Kakarougkas A, Meek K, Taucher-Scholz G, Lobrich M, Jeggo PA (2011) Factors determining DNA double-strand break repair pathway choice in G2 phase. EMBO J 30:1079–1092. https://doi.org/10.1038/emboj.2011.27

Shinozaki K, Yamaguchi-Shinozaki K, Seki M (2003) Regulatory network of gene expression in the drought and cold stress responses. Curr Opin Plant Biol 6:410–417

Shrivastav M, De Haro LP, Nickoloff JA (2008) Regulation of DNA double-strand break repair pathway choice. Cell Res 18:134–147. https://doi.org/10.1038/cr.2007.111

Singh SK, Roy S, Choudhury SR, Sengupta DN (2010) DNA repair and recombination in higher plants: insights from comparative genomics of Arabidopsis and rice. BMC Genomics 11:443. https://doi.org/10.1186/1471-2164-11-443

Slaughter A, Daniel X, Flors V, Luna E, Hohn B, Mauch-Mani B (2012) Descendants of primed Arabidopsis plants exhibit resistance to biotic stress. Plant Physiol 158:835–843. https://doi.org/10.1104/pp.111.191593

Steinert J, Schiml S, Puchta H (2016) Homology-based double-strand break-induced genome engineering in plants. Plant Cell Rep 35:1429–1438. https://doi.org/10.1007/s00299-016-1981-3

Steward N, Kusano T, Sano H (2000) Expression of ZmMET1, a gene encoding a DNA methyltransferase from maize, is associated not only with DNA replication in actively proliferating cells, but also with altered DNA methylation status in cold-stressed quiescent cells. Nucleic Acids Res 28:3250–3259

Strimbeck GR, Schaberg PG, Fossdal CG, Schroder WP, Kjellsen TD (2015) Extreme low temperature tolerance in woody plants. Front Plant Sci 6:884. https://doi.org/10.3389/fpls.2015.00884

Sung S, Amasino RM (2004) Vernalization and epigenetics: how plants remember winter. Curr Opin Plant Biol 7:4–10

Takeda S, Sugimoto K, Otsuki H, Hirochika H (1999) A 13-bp cis-regulatory element in the LTR promoter of the tobacco retrotransposon Tto1 is involved in responsiveness to tissue culture, wounding, methyl jasmonate and fungal elicitors. Plant J 18:383–393

Takeda S, Tadele Z, Hofmann I, Probst AV, Angelis KJ, Kaya H, Araki T, Mengiste T, Mittelsten Scheid O, Shibahara K, Scheel D, Paszkowski J (2004) BRU1, a novel link between responses to DNA damage and epigenetic gene silencing in Arabidopsis. Genes Dev 18:782–793. https://doi.org/10.1101/gad.295404

Tamang BG, Fukao T (2015) Plant adaptation to multiple stresses during submergence and following desubmergence. Int J Mol Sci 16:30164–30180. https://doi.org/10.3390/ijms161226226

Tessadori F, van Zanten M, Pavlova P, Clifton R, Pontvianne F, Snoek LB, Millenaar FF, Schulkes RK, van Driel R, Voesenek LA, Spillane C, Pikaard CS, Fransz P, Peeters AJ (2009) Phytochrome B and histone deacetylase 6 control light-induced chromatin compaction in *Arabidopsis thaliana*. PLoS Genet 5:e1000638. https://doi.org/10.1371/journal.pgen.1000638

Tiley GP, Burleigh JG (2015) The relationship of recombination rate, genome structure, and patterns of molecular evolution across angiosperms. BMC Evol Biol 15:194. https://doi.org/10.1186/s12862-015-0473-3

Tuteja N, Singh MB, Misra MK, Bhalla PL, Tuteja R (2001) Molecular mechanisms of DNA damage and repair: progress in plants. Crit Rev Biochem Mol Biol 36:337–397. https://doi.org/10.1080/20014091074219

Uthup TK, Ravindran M, Bini K, Thakurdas S (2011) Divergent DNA methylation patterns associated with abiotic stress in *Hevea brasiliensis*. Mol Plant 4:996–1013. https://doi.org/10.1093/mp/ssr039

Vaillant I, Schubert I, Tourmente S, Mathieu O (2006) MOM1 mediates DNA-methylation-independent silencing of repetitive sequences in Arabidopsis. EMBO Rep 7:1273–1278. https://doi.org/10.1038/sj.embor.7400791

Verhoeven KJ, Jansen JJ, van Dijk PJ, Biere A (2010) Stress-induced DNA methylation changes and their heritability in asexual dandelions. New Phytol 185:1108–1118. https://doi.org/10.1111/j.1469-8137.2009.03121.x

Wada Y, Miyamoto K, Kusano T, Sano H (2004) Association between up-regulation of stress-responsive genes and hypomethylation of genomic DNA in tobacco plants. Mol Genet Genomics 271:658–666. https://doi.org/10.1007/s00438-004-1018-4

Wei W, Ba Z, Gao M, Wu Y, Ma Y, Amiard S, White CI, Rendtlew Danielsen JM, Yang YG, Qi Y (2012) A role for small RNAs in DNA double-strand break repair. Cell 149:101–112. https://doi.org/10.1016/j.cell.2012.03.002

Yao Y, Kovalchuk I (2011) Abiotic stress leads to somatic and heritable changes in homologous recombination frequency, point mutation frequency and microsatellite stability in Arabidopsis plants. Mutat Res 707:61–66. https://doi.org/10.1016/j.mrfmmm.2010.12.013

Yao Y, Bilichak A, Golubov A, Kovalchuk I (2012a) ddm1 plants are sensitive to methyl methane sulfonate and NaCl stresses and are deficient in DNA repair. Plant Cell Rep 31:1549–1561. https://doi.org/10.1007/s00299-012-1269-1

Yao Y, Danna CH, Ausubel FM, Kovalchuk I (2012b) Perception of volatiles produced by UVC-irradiated plants alters the response to viral infection in naive neighboring plants. Plant Signal Behav 7:741–745. https://doi.org/10.4161/psb.20406

Yao Y, Bilichak A, Titov V, Golubov A, Kovalchuk I (2013a) Genome stability of Arabidopsis atm, ku80 and rad51b mutants: somatic and transgenerational responses to stress. Plant Cell Physiol 54:982–989. https://doi.org/10.1093/pcp/pct051

Yao Y, Kathiria P, Kovalchuk I (2013b) A systemic increase in the recombination frequency upon local infection of *Arabidopsis thaliana* plants with oilseed rape mosaic virus depends on plant age, the initial inoculum concentration and the time for virus replication. Front Plant Sci 4:61. https://doi.org/10.3389/fpls.2013.00061

Yoo S, Dynan WS (1998) Characterization of the RNA binding properties of Ku protein. Biochemistry 37:1336–1343. https://doi.org/10.1021/bi972100w

Youngson NA, Whitelaw E (2008) Transgenerational epigenetic effects. Annu Rev Genomics Hum Genet 9:233–257. https://doi.org/10.1146/annurev.genom.9.081307.164445

Zemach A, Li Y, Wayburn B, Ben-Meir H, Kiss V, Avivi Y, Kalchenko V, Jacobsen SE, Grafi G (2005) DDM1 binds Arabidopsis methyl-CpG binding domain proteins and affects their subnuclear localization. Plant Cell 17:1549–1558. https://doi.org/10.1105/tpc.105.031567

Zhou C, Zhang L, Duan J, Miki B, Wu K (2005) HISTONE DEACETYLASE19 is involved in jasmonic acid and ethylene signaling of pathogen response in Arabidopsis. Plant Cell 17:1196–1204. https://doi.org/10.1105/tpc.104.028514

Zhu JK (2009) Active DNA demethylation mediated by DNA glycosylases. Annu Rev Genet 43:143–166. https://doi.org/10.1146/annurev-genet-102108-134205

Zhu Q, Wani AA (2010) Histone modifications: crucial elements for damage response and chromatin restoration. J Cell Physiol 223:283–288. https://doi.org/10.1002/jcp.22060

Origin of Epigenetic Variation in Plants: Relationship with Genetic Variation and Potential Contribution to Plant Memory

Massimiliano Lauria and Vincenzo Rossi

Abstract Plants are sessile organisms that must cope with various environmental cues. Mechanisms allowing learning and memory are, therefore, essential for the plant life cycle. Various studies suggest that epigenetic-related mechanisms could play a key role in plant learning and memory. The importance of epigenetics in these processes mainly relies on its correlation with environment and the possibility to more easily revert the resulting phenotypic changes compared to those associated with the direct DNA sequence alteration. Nevertheless, an obvious cross talk between genetic and epigenetic variation occurs. This review discusses about the nature and origin of epigenetic variation, thus allowing to better understand the contribution of epigenetic variation to plant memory.

Keywords Epigenetic variation · Genetic variation · Plant memory · Environmental cues · Cytosine methylation

1 Introduction

In the mammalian brain, learning involves a series of molecular changes including regulation of gene transcription. Some of these changes are retained in post-mitotic cells throughout life and form the basis of the long-term memory. Stimuli which induce changes related to memory and learning can be of either endogenous or exogenous origins (Zovkic et al. 2013). Various studies provided evidence that epigenetic mechanisms, involving DNA methylation and chromatin modifications,

M. Lauria (✉)
Institute of Agricultural Biology and Biotechnology, Italian National Research Council, Milan, Italy
e-mail: lauria@ibba.cnr.it

V. Rossi (✉)
Council for Agricultural Research and Economics, Research Centre for Cereal and Industrial Crops, Bergamo, Italy
e-mail: vincenzo.rossi@crea.gov.it

play an essential role in learning and memory (Levenson and Sweatt 2005). The epigenetic machinery provides a strategy to regulate gene expression in response to specific signals during learning and occurs through the modulation of chromatin states. Subsequently, the same machinery allows the maintenance of the molecular changes during memory processes (Guan et al. 2015).

Plants are highly exposed to external cues, and due to their sessile nature, they have to find strategies to adapt to environmental-related changes. Hence, learning and memory are essential steps for plant response and adaption to environmental signals (Kinoshita and Seki 2014; Crisp et al. 2016). In this context, epigenetic variability may play a relevant role in plant adaption because these organisms appear particularly prone to transgenerational inheritance of epigenetic alterations (Quadrana and Colot 2016).

Understanding the nature and origin of epigenetic variability is, therefore, an important goal to unravel epigenetic-based mechanisms of learning and memory in plants and to develop approaches to improve plant adaption to biotic and abiotic stresses and climate changes. Although in many cases epigenetic variation is tightly linked to genetic variation (Pecinka et al. 2013), a particular type of epigenetic variation (i.e., the pure variation; Richards 2006) is conceptually considered as totally unrelated to the genetic context. This type of variation is of great interest as it may provide plants with an incredible level of plasticity, thus acting as a mechanism that permits plant to learn and memorize environmental experiences. Accordingly, in this review we will discuss recent advances in understanding the origin of epigenetic variability, with a particular focus on pure epigenetic variation and its potential contribution to plant memory.

2 Cytosine Methylation: A Brief Summary

The epigenetic information that is posed on a genome mainly relies on two specific marks: the methylation of cytosine (mC) and the posttranslational modifications of histones. In this review, we will focus principally on epigenetic variation associated with mC, and below a brief summary of mC regulation in plants is provided. For a more comprehensive description of the mC, readers can refer to various excellent reviews (Law and Jacobsen 2010; Pikaard and Mittelsten Scheid 2014; Matzke et al. 2015; Quadrana and Colot 2016; Martinez and Köhler 2017).

In plant, mC occurs at cytosine in all sequence contexts. These contexts are generally referred as symmetric (CG and CHG) and asymmetric (CHH, where H is A, T, or C). Establishing and maintaining these methylation features require specific enzymes, acting through different pathways. All cytosine contexts are methylated de novo via the RNA-directed DNA methylation (RdDM) pathway (Matzke et al. 2015). In this process, short-interfering 24nt RNAs (siRNAs) guide the de novo methyltransferase domains rearranged methyltransferase 2 (DRM2) enzyme to target sites (Cao and Jacobsen 2002a, b). Methylation at CG is maintained by methyltransferase 1 (MET1) during DNA replication. This enzyme recognizes the hemi-methylated CG sites and methylates the opposing strand (Finnegan et al. 1998) in cooperation with accessory proteins, such as variant in methylase 1 (VIM1;

Woo et al. 2007). Methylation at the CHG sites is maintained by the plant-specific enzymes chromomethylase 3 (CMT3) and, at least in part, chromomethylase 2 (CMT2; Lindroth et al. 2001; Stroud et al. 2013, 2014). In this case, both enzymes are targeted to CHG sites by dimethylation of lysine 9 of histone 3 (H3K9me2), which is recognized by the CHROMO domain of CMT2 and CMT3. The CMT2 enzyme also maintains methylation of CHH sites in H3K9me2-enriched regions (Zemach et al. 2013; Stroud et al. 2014).

Distribution of CG, CHG, and CHH is not uniform throughout the genome. In general, transposable elements (TEs) and other repeats are methylated in all cytosine contexts. Conversely, methylation in genes, especially those constitutively expressed, is mainly of the CG type and produces the typical gene body methylation profile: i.e., DNA methylation is mainly found in the central part of genes, and it is depleted/reduced at the 5'- and 3'-end regions (Zhang et al. 2006; Zilberman et al. 2006; Cokus et al. 2008; Lister et al. 2008). In methyltransferase mutant lines, loss of mC is associated with TEs reactivation, while there are not clear indications that loss of gene body methylation alters gene expression (Zhang et al. 2006; Bewick et al. 2016). Currently, gene body methylation is considered either a by-product, without functional consequence, of errant properties of enzymes that establish (i.e., CMT3) and maintain (i.e., MET1) CG methylation within genes or a homeostatic mechanism to stabilize gene expression (Bewick et al. 2016; Zilberman 2017). Moreover, although the molecular mechanisms that control mC distribution in plant genomes are generally conserved, a recent work has shown the existence of widespread variation of this epigenetic mark in plants (Niederhuth et al. 2016). Perhaps, the most strictly example is the loss of specific methyltransferase genes in some plant species, thus impairing their epigenetic landscape. Overall, these observations suggest that the nature and functional consequences of epigenetic variation varies between plant species.

3 Defining Epigenetic Variation

The first attempt to identify epigenetic mechanisms as possible source of epigenetic variation was likely proposed by Holliday in a seminal work published on the journal *Science* (1987). In the abstract of that article, the author stated: "Evidence from many sources shows that the control of gene expression in higher organisms is related to the methylation of cytosine in DNA, and that the pattern of methylation is inherited. Loss of methylation, which can result from DNA damage, will lead to heritable abnormalities in gene expression." The author also postulated that "Once it is accepted that epigenetic controls of the level of DNA transcription are heritable in cell lineages, then it follows that heritable defects or abnormalities in such controls are possible." With this article, Holliday introduced the concept of epimutations and epialleles and, hence, of epigenetic variation. Currently, the word "epimutation" is used as an equivalent of the DNA mutation. However, instead of permanently modifying the DNA sequence, epimutation alters the information that is posed on

Fig. 1 Different types of epigenetic variations on the basis of interaction with genetic variation. Schematic representation of the three kinds of epigenetic variation: obligatory, facilitated, and pure. The figure is based on the original definition provided by Richards (2006)

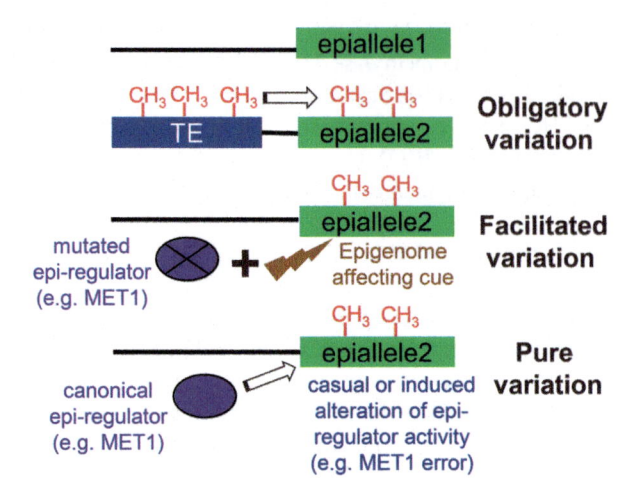

DNA sequence (i.e., DNA methylation of cytosine) or its associated proteins (i.e., histones). These alterations can be inherited both mitotically and meiotically, but they have a higher potential than DNA mutation to be reverted to the original state. If an epigenetic mutation generates alternative and heritable states of gene expression, then we can say that an "epiallele" has been generated.

Thirty years after Holliday's article, many important advances have been made in the field of epigenetics, and this has been possible, thanks to the use of model species and development of genetic tools and novel technologies that allow for direct assessment of the epigenome features of an organism at high resolution (Law and Jacobsen 2010; Pikaard and Mittelsten Scheid 2014; Quadrana and Colot 2016). However, some aspects of epigenetic variation and the relevance of its functional consequence remain unclear. One specific issue concerning epigenetic variation is to establish at which extent it is related to genetic diversity. This is an essential aspect to fully understand (1) the mechanisms of epigenetic variation origin, (2) to what extent and through what mechanism epigenetic variation is associated with phenotypic changes, and (3) the strategy to be used for exploring and possibly manipulating epigenetic variation.

One of the first classification of the possible relationship between genetic and epigenetic variation was proposed by Richards (2006; Fig. 1). Three levels of interaction between epigenetic states and their genotypic context were proposed; these levels of interactions are classified as obligatory, facilitated, and pure epigenetic variation. A more recent redefinition of these interactions was proposed by Eichten et al. (2014) on the basis of their stability and of the *cis/trans* influence of genetic over epigenetic variation. Furthermore, Pencika et al. (2013) reconsidered the three classes proposed by Richards (2006) in the view that genetic variation is likely the major source of epigenetic variation origin; a concept that was also introduced by Meagher (2010). In this review, for simplicity, we will refer to the original classification proposed by Richards (2006; Fig. 1).

Obligatory epigenetic variation occurs when epigenetic variation is completely dependent on underlying genetic variation (Fig. 1). One of the most effective examples of obligatory variation is the depletion of methylable cytosine sites as result of either single nucleotide polymorphisms (SNPs) or sequence deletion (Meagher 2010). DNA variation can also impact nucleosomes positioning, since this process is strongly determined by DNA sequence. Altering nucleosome phasing may have also a great impact on gene expression if the change involves the binding sites for transcriptional activators (Meagher 2010). Sequence variation generated by TEs has also the potential to generate obligatory epigenetic changes. TEs are direct targets of silencing mechanisms mediated by siRNAs to limit their negative effect on genome stability (Eichten et al. 2012; Li et al. 2015). The result is that TEs are characterized by DNA hypermethylation and accumulation of heterochromatin-related histone marks (e.g., H3K9me2), thus determining the formation of a tight chromatin structure, very poorly accessible to the RNA polymerase II transcriptional machinery. The possible spreading of these silencing epigenetic marks to neighbor coding sequences can lead to a consistent repressive transcriptional state. In this case, obligatory epigenetic variation can be simply considered as a direct consequence of DNA features, and the resulting phenotypes can be studied within a population using classical genetic approaches.

The same genetic features that lead to obligatory epigenetic states can, under different circumstances, produce epigenetic states that are less predictable (Fig. 1). The resulting locus can be activated or repressed in a more probabilistic manner. These manifestations show a metastable nature, and they belong under the class of the facilitated epigenetic variation, because of the stochastic variation of the epigenetic status associated with a DNA sequence change (Richards 2006). A simplified example of facilitated variation could be a genotype bearing a mutation in one epigenetic regulator (epi-regulator) that, under concomitant exposure to external factors like stresses, facilitated the formation and the maintenance (i.e., the inheritance after the depletion of the external stimulus) of an epiallele.

Finally, the last class of epigenetic variation is defined as pure and includes examples of epigenetic states that are completely independent from DNA sequences (Fig. 1). This type of epigenetic variation occurs with mechanisms resembling those responsible for genetic variation. For example, similarly to DNA polymerase I, it is possible that MET1 makes errors in methylating cytosines during DNA replication and that these epimutations are inherited through mitosis and/or meiosis.

Since obligatory and facilitated epigenetic variation originate from genetic components, their kind of contribution to memory is similar to that of genetic variation. In particular, like genetic variation, these types of epigenetic variation lack of some important features predicted to be essential for memory: (1) the ability to "learn" and store as a molecular signature a particular environmental signals in a not stochastic manner and (2) the possibility to easily revert the resulting phenotypic changes. These features are instead intrinsic in pure epigenetic variation, which is the only kind of epigenetic variation that could contribute to plant memory.

It is, however, important to note that the concept of pure epigenetic variation relies on our ability to recognize the epigenetic potential of a given DNA sequence in terms of establishing and maintaining a specific epigenetic state. For example, it could be difficult to establish a link between a genetic determinant and an epigenetic state, because the former could be located at a great distance with respect to the latter. Even if the two determinants are closely located, we may simply miss their connections because of our still limited knowledge about how a specific DNA sequence may affect the epigenetic information. An example is provided by the recent findings of Gouil and Baulcombe (2016), which provide the novel indication about the importance of the DNA sequence "sub-context" for the methylome in different plant species. The authors found that, in in heterochromatin region of four plant species, CCG methylation is 20–50% lower than CAG and CTG. Moreover, when CG sub-contexts is considered, it was found that in tomato and *Arabidopsis* CGT methylation is lower compared to other sub-contexts, while in rice CGA and CGT are generally more methylated than CGG and CGC, and in maize there is not an apparent link between CG methylation and sub-context. Finally, it is important to point out that without availability of a complete genome sequence of individuals with distinct epialleles, we may simply be unable to identify the potential genetic determinant of these epialleles. For example, when an environmental-related stimulus is associated with the formation of epialleles, only after a careful analysis of the complete genome sequences of individuals we can establish if different epialleles are due to pure or obligatory/facilitated epigenetic variation, because stresses often induce TEs-mediated insertion/deletion/recombination events, which, in turn, may direct epigenetic changes (Pecinka et al. 2013). In conclusion, despite several examples of natural and induced epialleles that have been identified in plants, to the best of our knowledge, none of them have been unambiguously demonstrated the fully independence by genetic determinants. The only documented exception could arise from experimental strategies specifically developed to characterize, at a genome-wide level, pure genetic variation in *Arabidopsis* (*Arabidopsis thaliana*; see below).

4 Examples of Different Kinds of Plant Epigenetic Variation on the Basis of Its Interaction with Genetic Determinants

Various examples of obligatory epigenetic variation have been reported in plants (Pecinka et al. 2013; Eichten et al. 2014). Among the best characterized is the *FLOWERING WAGENINGEN A* (*FWA*) gene containing a DNA hypermethylated *SHORT INTERSPERSED ELEMENT* (*SINE*) in the 5' regulatory region. During normal development, *FWA* gene is specifically expressed in endosperm and in an imprinted manner (Choi et al. 2002). It was reported that *Arabidopsis* mutants in the chromatin-remodeling factor *DECREASED IN DNA METHYLATION1* (*DDM1*) exhibit loss of methylation at *SINE*, thus inducing ectopic *FWA* expression during vegetative growth and a delay of flowering (Soppe et al. 2000; Kinoshita et al. 2007). Interestingly, analysis of *FWA* in several *Arabidopsis* species has revealed interspecific variation in the *SINE* structure, indicating that a higher number of *SINE* tandem repeats ensure stronger *FWA* silencing (Fujimoto et al. 2008). Therefore, it is evident that genetic variability in an epigenetic regulator (i.e., *DDM1*) as well as a structural variation (i.e., the numbers of tandem repeats within *SINE*) is the direct determinant of the epigenetic variations.

An interesting example of facilitated epigenetic variation is provided by a recent study that has characterized the metastable nature of important agronomic traits in the oil palm fruit (Ong-Abdullah et al. 2015). In some parts of the world, this plant species is widely used for oil production. Because of the high oil demand, genetic strategies have been employed to improve yield. One of these strategies is based on micropropagation, through cell culture, of immature apex leaf tissue of genotypes that are known to increase oil yield by up to 30% (Corley and Law 1997). However, this procedure generates a subset of plants characterized by a mantled phenotype, which leads to important yield loss as a result of sterile parthenocarpic flowers with abortive fruit (Corley 1986). The mantled phenotype exhibits a metastable nature because it can shift from a mutant to wild-type state and vice versa. In addition, this shift occurs in a non-Mendelian manner. Therefore, mantled phenotype has been considered as an example of epigenetic variation (Jaligot et al. 2000). The recent study of Ong-Abdullah et al. (2015) clarified the epigenetic bases of this phenomenon. The authors utilized an epigenome-wide association study (i.e., analysis of mC polymorphisms instead of classical DNA polymorphisms) to identify loci associated with mantled phenotype. It was found that mantled and wild-type plants differ for only one genomic region, which is differentially methylated in a stable manner. This region lies within the 35 kb intron 5 of *EgDEF1* gene: an ortholog of the B class MADS box transcription factor genes *DEFICIENS* of *Antirrhinum majus* and *APETALA3* of *Arabidopsis*. Interestingly, the differentially methylated region overlaps with a repetitive element that exhibits homology to rice (*Oryza sativa*) Karma *LONG INTERSPERSED ELEMENTS* (*LINEs*) element. In mantled plants, *Karma* shows a reduced level of CHG methylation, which modifies the ability of *EgDEF1* to produce a correct transcript and, therefore, a functional protein. Demethylation of *Karma* element in oil palm likely depends on the artificial procedure (micropropagation) used to generate these plants, which is known to induce alteration of the epigenome (Tanurdzic et al. 2008; Stroud et al. 2013). Nevertheless, it is worth noting that these phenomena are frequent in the wild, especially in plants that are characterized by a pervasive presence of TEs, and that may be exposed to a wide range of stressful conditions.

A possible example of pure epigenetic variation is the natural epiallele identified in *Linaria vulgaris*. In this plant, a region responsible for a change from bilateral to radial (i.e., peloric) floral symmetry was mapped to the *cis*-regulatory region of the *CYCLOIDEA*-like (*Lcyc*) gene (Cubas et al. 1999). It was shown that differences between epialleles are due to DNA hypermethylation and silencing of *Lcyc* in peloric with respect to bilateral flowers. In addition, phenotypic reversion from peloric to bilateral flowers correlates with reduced mC level. Furthermore, the simultaneous presence of peloric, intermediate, and revertant flowers within a single plant strongly supports the conclusion that *Lcyc* is an epiallele. Despite searching, it has not yet been demonstrated that either the genome sequence of peloric and bilateral flowers is truly identical or that these events are facilitated by the presence of specific DNA sequences which have not been observed. Even in a single individual, examples of genetic mosaicism have been reported as the result of somatic transposition, and it is intriguingly that, in mammalian brain, this phenomenon is likely related to processes linked to memory (Richardson et al. 2014). Therefore, without availability of genome sequence of flower cells exhibiting different phenotypes, it is not possible to rule out a possible contribution of genetic variation as determinants of *Lcyc* epialleles.

To classify the true nature of epigenetic variation, other aspects, different to availability of the genome sequences, must be considered. Indeed, an extensive behavior analysis of the epiallele performed over the time is also required to define the real nature of the variation. In this respect, the *B* locus of maize (*Zea mays*) represents an excellent example. In maize, the *b1* gene encodes a transcription factors that activate the biosynthesis of flavonoid pigments (Coe 1959; Goff et al. 1992). The *B-Intense* (*B-I*) allele confers high pigmentation to the plant tissues of maize, and it can spontaneously shift to a silenced state, namely, *B'*. The *B-I* and *B'* states represent the basic elements of the first epigenetic phenomenon described so far. This phenomenon, deemed paramutation, is an epigenetic mechanism where one allele (defined paramutagenic allele) causes a heritable change in the expression of a homologous allele (defined paramutable allele; Hollick 2010). However, *B-I* and *B'* represent a particular, although not unique, example of paramutation, because the paramutagenic allele is a direct derivate of the paramutable allele. To unravel the molecular basis of *B-I* and *B'* difference, an extensive genetic and molecular characterization of this locus has been performed by means of restriction map analyses (Patterson et al. 1993; Stam et al. 2002a, b). These analyses did not find clear evidence of differences between the DNA sequences of the two alleles. Moreover, these studies identified a genomic region located 100 kb far from the B1 coding region and that is required for the transition from *B-I* to *B'* (Stam et al. 2002a). This region contains seven tandem repeats that are the key determinants for the epigenetic behavior observed at the *B-I* allele. Therefore, on the basis of these results, *B-I* allele may represents either an example of pure epigenetic variation, because *B-I* and *B'* do not exhibit evident genetic differences, or an example of facilitated epigenetic variation, because tandem repeats induce the *B-I* to *B'* transition.

However, extensive analysis of *B'* progeny over generations showed that, while the frequency of transition from *B-I* to *B'* ranges from 1% to 10%, the opposite (i.e., reversion from *B'* to *B-I*) has never been observed (Chandler and Alleman 2008). Hollick (2010) proposed that such behavior should lead to the extinction of *B-I* in an open pollinated population and that its persistence may be explained by strong human selection of this locus. Therefore, this example raises the question if the behavior of *B-I* represents the natural course of an allele that is the subject of an obligatory epigenetic event.

5 Experimental Approaches to Study Pure Epigenetic Variation

To minimize the confounding effect of genetic over the epigenetic variation and to identify examples of pure epigenetic variation, different strategies have been adopted. All of them are based on the use of individuals characterized by limited genetic variation (ideally by its absence). In this respect, plants provide a great opportunity to disentangle the characteristic of this phenomenon, as highly genetically identical populations can be easily created. Indeed, a majority of plants reproduce by selfing or asexually (Verhoeven et al. 2010, 2011). Moreover, tools to induce highly identical genetic plants are available, such as the haploid-diploid doubling process (Hauben et al. 2009).

The strategies used to study pure epigenetic variation can be classified as "induced" or "natural." Induced approaches take advantage of experimental procedures that influence individual epigenomes. For example, chemical demethylation agents, such as 5-azacytidine and zebularine, can be used to alter mC patterns of individuals (Bossdorf et al. 2010; Baubec et al. 2013; Xu et al. 2016). Likewise, plants generated by tissue propagation are well known to produce altered mC patterns that are transgenerationally stable, at least for some generations (Stroud et al. 2013; Stelpflug et al. 2014). Interestingly, the targets of the induced variations are quite random. Stroud et al. (2013) observed that regenerated rice plants, which share the same parent, exhibit distinct mC profiles, evidencing that the tissue culture procedure induces stochastic changes. Similarly, studies conducted with chemical treatments produce plant populations that are characterized by distinct phenotypes (Bossdorf et al. 2010) and differences in the mC pattern (Xu et al. 2016). The randomness in producing epigenetic changes observed in these studies, along with the heritability of phenotypes and epigenetic marks, shows that these strategies may be effective to create populations characterized by distinct epigenotypes and, therefore, useful for testing the functional and phenotypic effect of epigenetic variation with large-scale studies.

Plant methylomes can also be altered using epi-regulator mutants, such as those involved in maintenance of mC patterns through mitosis and meiosis. In these respects, the development of *Arabidopsis* epigenetic recombinant inbred lines (epiRIL) has provided incredible tools for studying several expects of plant epigenetics in terms of functional consequences and hereditability. These lines are highly variable at the epigenetic level but nearly identical at the DNA sequence level. They were achieved using *MET1* and *DDM1* mutant alleles (Johannes et al. 2009; Reinders et al. 2009). The scheme for achieving epiRIL is illustrated in Fig. 2. With epiRILs is possible to obtain various genetically identical lines, each of them bearing specific regions of the genome that are demethylated in both chromosomes. Importantly, epiRIL populations contain extensive phenotypic variation that was stable throughout each population for numerous traits, such as time to flowering, plant height, and fruit size (Johannes et al. 2009; Reinders et al. 2009; Fig. 2). One

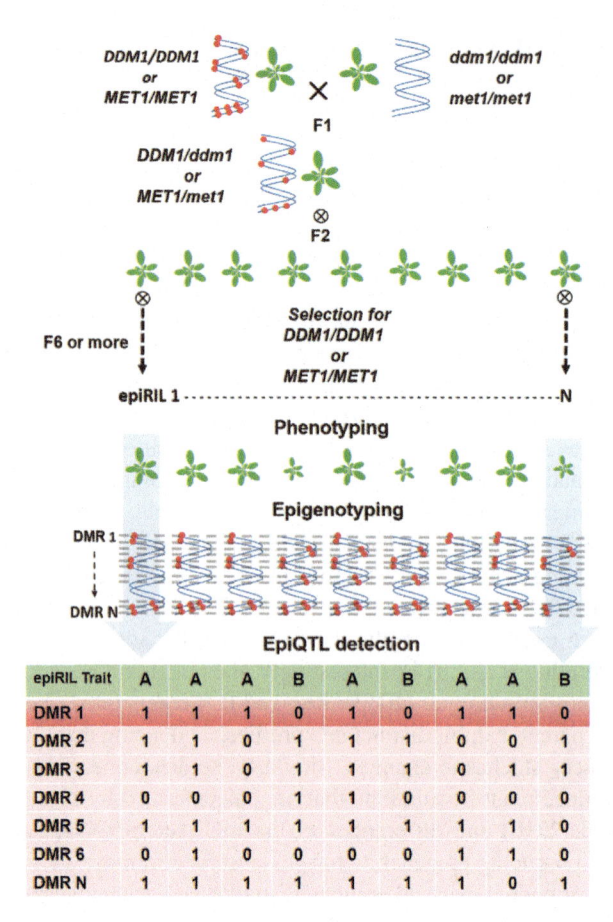

Fig. 2 Strategy for production and analysis of epiRILs. Both *MET1* and *DDM1* encode for proteins that affect the propagation of mC patterns. For obtaining epiRILs, each of the two mutants, the mutant line, was crossed with a near-isogenic wild-type line to generate F1 hybrids. Remarkably, remethylation of *met1* and *ddm1* chromosomes does not occur immediately upon introduction to wild-type chromosomes in the F1 hybrid; instead, remethylation occurs gradually over generational time. Subsequently, segregating populations of plants were screened for the presence of wild-type alleles of either *MET1* or *DDM1* and propagated by single seed descent for multiple generations to increase homozygosity. This process produces a population of epiRILs, and each epiRIL essentially contains a mosaic epigenome derived from either wild-type and *met1* mutant or wild-type and *ddm1* mutant. A simplified example of the epiRILs analysis is illustrated in the figure. The analysis is based on a combination of epigenotyping (DMRs identification) and phenotyping, followed by the application of quantitative genetics approaches, like the genome-wide association study (GWAS), and for achieving epiQTLs associated with variation of specific traits

problem with epiRILs is that a degree of genetic variation, due to TEs mobilization in response to alteration of *MET1* or *DDM1* function, was observed through distinct lines (Johannes et al. 2009; Reinders et al. 2009; Mirouze et al. 2009). Therefore, in order to truly identify examples of pure epigenetic variation, complete knowledge of the genome sequence of individuals under study is required. Although, for some plant species, such as *Arabidopsis*, this information can reasonably be obtained, for other species, which lack of a reference genome sequence or that have a large and complex genome, this information is difficult to acquire, due to both technical and economic issues. For these species, the independence between genetic and epigenetic effect can be only statistically evaluated on the basis of the different rate between the observed genetic and epigenetic variability. In addition, an epiRIL strategy could be challenging to develop for some of the major crops. For example, in maize, likely due to the presence of duplicate and functional redundant *MET1* and *DDM1* copies within its genome, all the attempts to produce epiRILs have been not successful (Li et al. 2014). Interestingly, even in rice, where only a single copy of *MET1*-like allele is present, the production of homozygous *met1* mutant plants failed (Hu et al. 2014). The different ability of plant species to support the lack of DNA methylation may be related to their different TEs content and to higher frequency of genes located close to TEs with respect to *Arabidopsis* (Li et al. 2014). These species have likely developed mechanisms to protect genes, embedded into a TEs rich genome, from their silencing, due to the spreading of heterochromatin-inducing and TE-related epigenetic marks (Lisch 2012). In addition, these species have obviously developed strategies to silence TEs activities, thus avoiding excessive genome instability (Lisch and Bennetzen 2011). Therefore, loss of pivotal epi-regulator-like *MET1* and *DDM1* may be not compatible with viability because of the deregulation of TE activity.

A final strategy to study the extent and role of pure epigenetic variation is to use the so-called epimutation accumulation lines (epiMAs). Basically, this approach can be considered natural as it relies on the fact that spontaneous epimutations can arise as result of mechanisms that perpetuate epigenetic states from one generation to the next (Becker et al. 2011; Schmitz et al. 2011). These studies used a set of *Arabidopsis* plants, generated from a lone founder by single seed descent growth for 30 generations. These plants were previously used to assess the DNA mutation rate (Ossowski et al. 2010), thus allowing estimation of mC changes unlinked to genetic variation. The strategy used for epiMAs analysis is illustrated in Fig. 3.

Although all the abovementioned studies represent experiments that take advantage of artificial laboratory conditions, which do not truly reflect what happens in the "real world," they still represent events that may also occur in wild. For example, accumulation of DNA mutations in natural populations could affect functions of epi-regulators, thus increasing the epimutations rate and, consequently, the population's epigenetic landscape. Indeed, when 89 different *Arabidopsis* ecotypes were used to screen for mC natural variation at the centromeric 180-base-pair sequence, one of the result was the identification of a new epi-regulator, the VIM1 methyl-binding protein (Woo et al. 2007). The *Arabidopsis* Bor-4 accession carries a *vim1* mutation, and the centromeric repeat shows a level of hypomethylation comparable to that observed in *ddm1* mutants. A further example is provided by the study of Becker et al. (2011). Among epiMA lines used to evaluate the extent of mC variation, one strain had 40% more differentially methylated sites in comparison with other strains. This strain has a non-synonymous change in the amino acid

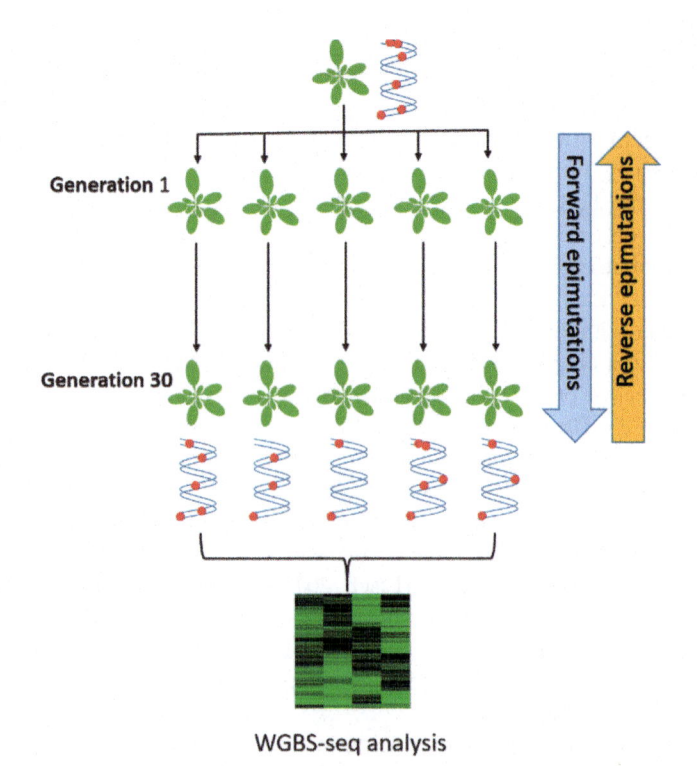

Fig. 3 Strategy for production and analysis of epiMAs. To achieve epiMAs, *Arabidopsis* plants were generated from a lone founder by single seed descent growth for 30 generations. The DNA mutation rate is assessed through application of whole-genome next-generation sequencing (NGS). Whole-epigenome NGS (e.g., whole genome bisulfite sequencing, WGBS-seq) is then applied to estimate single methylated cytosine changes and DMRs. Following mC variation through generations also allows to identify the precise rate of forward and reverse epimutations

sequence of MATERNAL EFFECT EMBRYO ARREST57 (MEE57) protein, which is related to MET1 function. Although, the study did not clarify how this allelic variation contributes to the exceptional level of mC variation observed in that epiMA line, this example, along with others reported in literature (Shen et al. 2014; Dubin et al. 2015), point out how the presence of epi-regulators genetic variants in natural populations may significantly influence the epimutations rate, as well as the real nature of epigenetic variation origins. In this context, since the complete elimination of *MET1* and *DDM1* function seems to be very challenging in some crops (see above), the identification of their genetic variants, as well as of variants of other important epi-regulators, in natural accessions/ecotypes could be a strategy to circumvent difficulties to "create" epimutations through the classical epiRIL approach.

Finally, it is worth to mention that epigenetic alterations induced by environmental stresses have been long considered as a strategy used by plants to increase their fitness in wild. A large body of evidence in the last decades has tried to provide support to this phenomenon and to establish whether and how this natural plant strategy could be coopted to produce and/or analyze epimutations in genetically identical individuals. Nonetheless, this specific aspect is the subject of various chapter of the present book; therefore, we will not discuss it further.

6 Functional Consequence of Pure Epigenetic Variation

Currently, a number of studies made use of genetically identical populations, thus allowing to better clarify the link between genetic and epigenetic variation (Hauben et al. 2009; Verhoeven et al. 2010, 2011; Li et al. 2015; Xu et al. 2016). In this section, we will refer to two kinds of exemplificative studies. These studies are of particular relevance because they completely fulfill one essential prerequisite for identifying pure epigenetic variation: the availability of the genome sequence to establish whether genetic variants, or lack thereof, influence epigenetic variability.

The first kind of study employs *Arabidopsis* epiMA (Becker et al. 2011; Schmitz et al. 2011; Fig. 3). The results show that, under controlled greenhouse conditions, the number of spontaneous single cytosine epimutations identified is four to five orders of magnitude greater than that of DNA sequence mutations. The majorities of these changes are methylation polymorphisms at single CG sites and mainly occur in gene coding regions. Despite abundance of single mC variation among individuals, the impact on gene expression is apparently low, if any. As previously mentioned, the lack of correlation between such abundant mC variation and gene expression may reflect the nonfunctional role of gene body methylation (Bewick et al. 2016). Alternatively, if gene body methylation is functionally relevant, its function could be not linked to gene expression (Takuno and Gaut 2012), which, hence, is not altered following variation of gene body methylation. A further important outcome from these studies is that differentially methylated regions (DMRs; i.e., stretches of concurrent cytosines displaying variation between samples) are less frequent than mutation at single cytosine methylation and is similar to the genetic mutations rate. Although DMRs are known to alter gene expression only for a small number of genes that they overlap, a general negative correlation between DMRs and expression was observed (Becker et al. 2011; Schmitz et al. 2011). Nevertheless, it is worth noting that although these studies have quantified the extent of mC variation and the rate of forward and reverse epimutations, they leave unresolved the impact of mC variation on these plants.

A second kind of study that provided relevant information about effect of pure epigenetic variation on the phenotype is represented by the analysis of the epiRILs using quantitative genetics approaches (Fig. 2). Many traits have been evaluated and found to vary heritably in the population of over 500 *ddm1*-derived epiRILs (Johannes et al. 2009; Roux et al. 2011; Cortijo et al. 2014; Kooke et al. 2015). Importantly, several quantitative trait locus (QTL) intervals corresponding to stable DMRs were identified for various traits, and they account for 60–90% of the heritability for two complex traits, such as the flowering time and primary root length. Many of these DMRs are also variable in *Arabidopsis* natural populations. In addition, the so-called epiQTLs are reproducible and can be subjected to artificial selection. This suggests that epiallelic variation contributes to heritable differences in complex traits in nature and independently of DNA sequence changes. Whole genome sequencing of epiRILs ruled out that genetic variation found in epiRILs was associated with the epiQTLs identified. However, it is believed that TEs can become mobilized in epiRILs due to the relaxation of the repressive methylation found at genomic elements. TEs and structural variation remain common factors in natural populations; it is not possible to unambiguously establish that the epigenetic variation associated with epiQTLs and phenotypic changes is a pure variation. Indeed, TEs mobilization may have somehow facilitated or even directed the formation of DMRs, possibly through mediation of interactions between long-distantly located genomic regions. Relevance of these interactions for plant chromatin's three-dimensional organization and its effect on genome activity has been recently demonstrated (Feng et al. 2014; Grob et al. 2014; Wang et al. 2015). Therefore, interactions between genomic regions may account for distant *trans* acting regulation of epigenetic changes.

7 Concluding Remarks

One of the major conclusions from the present review is that the contribution of pure epigenetic variation to the total epigenetic variation is still unclear. Nevertheless, it seems evident that a tight relationship between genetic and epigenetic variation exists and that genetic determinants likely contribute the majority of total epigenetic variation observed to date (Pecinka et al. 2013; Seymour and Becker 2017). An important support to this conclusion arises from a recent study that analyzed the methylome and transcriptome of more than 1000 *Arabidopsis* accessions (Kawakatsu et al. 2016). This work evidenced that epigenetic variation often originates from genetic variation (SNPs), structural variation (TEs), and trans-acting variants. In particular, physical maps for nine of the most diverse genomes revealed how TEs and other structural variants shape the epigenome, with dramatic effects on immunity genes.

If genetic variation is the major determinant of epigenetic variation, this raises the question of what kind of contribute the latter provides to plant memory, since it will depend only by pure epigenetic variation. By assuming that pure epigenetic variation mainly reflects stochastic errors of machineries involved in the propagation of epigenetic states, then these errors could be positively selected within a population because of their beneficial effect on the individual fitness. In this case, however, like for genetic variation, it would be the selection of a stochastic event and not the memory of a precise event, the mechanism leading to the transgenerational inheritance of phenotypes. Therefore, pure epigenetic variation can play a role in memory only when it is directed by specific environmental conditions at specific loci or genomic regions, thus allowing plants to acquire novel expressional states that are useful to their fitness, as well as for fitness of their progeny. Interestingly, a couple of studies indicate that different stresses target epigenetic variation in distinct genomic regions (Seymour et al. 2014; Wibowo et al. 2016). This observation suggests that during evolution, some regions of the genome might have been selected to maintain a sort of epigenetic plasticity, perhaps as a part of a learning mechanism, which is useful for dynamic adaptation to environmental fluctuation. However, to be effective as a memory mechanism, it is essential that these environmental-induced epigenetic variations are maintained within a population through generations. Materials characterized by limited genetic variation, such the epiRIL or inbreed lines, are essential tools to analyze this aspect in distinct plant species. For example, these materials could be exposed to stressful growth conditions to evaluate (1) the grade and rate of adaptation that is associated with epigenetic variation and (2) to what extent this variation is transgenerationally inherited. In this context, a series of recent studies have explored the ability of genetically identical plant populations to acquire novel environmental-induced epigenetic states, but they found little evidence that these alterations can be stably inherited (Eichten and Springer 2015; Hagmann et al. 2015; Secco et al. 2015; Wibowo et al. 2016). In addition, Wibowo et al. (2016) showed that at least two rounds of recurrent stress exposure are required for permitting transgenerational inheritance of induced epigenetic changes, and in the absence of a renewed stimulus, these epigenetic alterations are reset in subsequent generations. These observations suggest that the role of pure epigenetic variation in plant memory, in terms of transmission of an acquired experience, is minimal. Nevertheless, an alternative hypothesis should be also considered. Plants must have efficient mechanisms for a rapid response to stress (Crisp et al. 2016), which allow them to be highly flexible to various and fast environmental changes. In addition, the processes required for memory have a cost in terms of energy. Therefore, it could be predicted that the acquisition of an "epigenetic memory" only occur after a prolonged exposition to stimuli. Perhaps, this will also avoid posing on the offspring to information that may result unnecessary, or even deleterious, if the offspring will not experience the same parental stress that originally generated an epigenetic change (Lamke and Baurle 2017). It will be of great interest to test this hypothesis using epiRILs and epiMAs by applying different and prolonged stresses, possibly using plants grown in the field (i.e., in the "real world") and sampled under the prevailing environmental conditions.

As a final consideration, since most of epigenetic variation is apparently of genetic dependent and it is unclear if pure epigenetic variation really exists, we propose to modify accordingly the definition of epigenetics. The aim is to include in the definition the information about the "elusive" origin of epigenetic variability. Specifically, we propose to slightly change the sentence, often employed in epigenetics definition, which states that "epigenetics represents alteration of gene and genome activity not associated with variation of the DNA sequence." A more appropriate definition may be the following: "epigenetics is the discipline that studies the inheritance of the information stored within chromatin (hence not linked to the sequence of the DNA double helix) and that persists after that the stimulus inducing this information has been removed, independently by the nature of mechanisms mediating its origin (hence the exogenous or endogenous stimulus can induce the epigenetic variation, independently that this occurs by means of mechanisms directly or indirectly related or fully unrelated to genetic variation)."

Acknowledgments We thank Steve Eichten for critical reading of the manuscript and useful suggestions. The work regarding epigenetic studies in Massimiliano Lauria and Vincenzo Rossi laboratories is principally supported by special grants from the Italian Ministry of Education, University and Research (MIUR) and the National Research Council of Italy (CNR) for the Epigenomics Flagship Project (EPIGEN).

References

Baubec T, Ivánek R, Lienert F, Schübeler D (2013) Methylation-dependent and -independent genomic targeting principles of the MBD protein family. Cell 153:480–492
Becker C, Hagmann J, Müller J, Koenig D, Stegle O, Borgwardt K, Weigel D (2011) Spontaneous epigenetic variation in the Arabidopsis thaliana methylome. Nature 480:245–249
Bewick AJ, Ji L, Niederhuth CE, Willing EM, Hofmeister BT, Shi X, Wang L, Lu Z, Rohr NA, Hartwig B, Kiefer C, Deal RB, Schmutz J, Grimwood J, Stroud H, Jacobsen SE, Schneeberger K, Zhang X, Schmitz RJ (2016) On the origin and evolutionary consequences of gene body DNA methylation. Proc Natl Acad Sci USA 113:9111–9116
Bossdorf O, Arcuri D, Richards CL, Pigliucci M (2010) Experimental alteration of DNA methylation affects the phenotypic plasticity of ecologically relevant traits in Arabidopsis thaliana. Evol Ecol 24:541–553
Cao X, Jacobsen SE (2002a) Locus-specific control of asymmetric and CpNpG methylation by the DRM and CMT3 methyltransferase genes. Proc Natl Acad Sci USA 4:16491–16498
Cao X, Jacobsen SE (2002b) Role of the Arabidopsis DRM methyltransferases in *de novo* DNA methylation and gene silencing. Curr Biol 12:1138–1144
Chandler V, Alleman M (2008) Paramutation: epigenetic instructions passed across generations. Genetics 178:1839–1844
Choi Y, Gehring M, Johnson L, Hannon M, Harada JJ, Goldberg RB, Jacobsen SE, Fischer RL (2002) DEMETER, a DNA glycosylase domain protein, is required for endosperm gene imprinting and seed viability in Arabidopsis. Cell 110:33–42
Coe EH (1959) A regular and continuing conversion-type phenomenon at the b locus in maize. Proc Natl Acad Sci USA 45:828–832
Cokus SJ, Feng S, Zhang X, Chen Z, Merriman B, Haudenschild CD, Pradhan S, Nelson SF, Pellegrini M, Jacobsen SE (2008) Shotgun bisulphite sequencing of the Arabidopsis genome reveals DNA methylation patterning. Nature 452:215–219
Corley RHV (1986) Oil palm. In: Monselise SP (ed) CRC handbook of fruit set and development. CRC Press, Boca Raton, pp 253–259
Corley RHV, Law IH (1997) The future for oil palm clones. In: Pushparajah E (ed) Plantation management for the 21st century. Incorp Soc Planters, Kuala Lumpur, pp 279–289
Cortijo S, Wardenaar R, Colomé-Tatché M, Gilly A, Etcheverry M, Labadie K, Caillieux E, Hospital F, Aury JM, Wincker P, Roudier F, Jansen RC, Colot V, Johannes F (2014) Mapping the epigenetic basis of complex traits. Science 343:1145–1148
Crisp PA, Ganguly D, Eichten SR, Borevitz JO, Pogson BJ (2016) Reconsidering plant memory: intersections between stress recovery, RNA turnover, and epigenetics. Sci Adv 2:e1501340
Cubas P, Vincent C, Coen E (1999) An epigenetic mutation responsible for natural variation in floral symmetry. Nature 401:157–161
Dubin MJ, Zhang P, Meng D, Remigereau MS, Osborne EJ, Paolo Casale F, Drewe P, Kahles A, Jean G, Vilhjálmsson B, Jagoda J, Irez S, Voronin V, Song Q, Long Q, Rätsch G, Stegle O,

Clark RM, Nordborg M (2015) DNA methylation in Arabidopsis has a genetic basis and shows evidence of local adaptation. eLife 4:e05255

Eichten SR, Springer NM (2015) Minimal evidence for consistent changes in maize DNA methylation patterns following environmental stress. Front Plant Sci 6:308

Eichten SR, Ellis NA, Makarevitch I, Yeh CT, Gent JI, Guo L, McGinnis KM, Zhang X, Schnable PS, Vaughn MW, Dawe RK, Springer NM (2012) Spreading of heterochromatin is limited to specific families of maize retrotransposons. PLoS Genet 8:e1003127

Eichten SR, Schmitz RJ, Springer NM (2014) Epigenetics: beyond chromatin modifications and complex genetic regulation. Plant Physiol 165:933–947

Feng S, Cokus SJ, Schubert V, Zhai J, Pellegrini M, Jacobsen SE (2014) Genome-wide Hi-C analyses in wild-type and mutants reveal high-resolution chromatin interactions in Arabidopsis. Mol Cell 55:694–707

Finnegan EJ, Genger RK, Peacock WJ, Dennis ES (1998) DNA methylation in plants. Annu Rev Plant Physiol Plant Mol Biol 49:223–247

Fujimoto R, Kinoshita Y, Kawabe A, Kinoshita T, Takashima K, Nordborg M, Nasrallah ME, Shimizu KK, Kudoh H, Kakutani T (2008) Evolution and control of imprinted FWA genes in the genus Arabidopsis. PLoS Genet 4:e1000048

Goff SA, Cone KC, Chandler VL (1992) Functional analysis of the transcriptional activator encoded by the maize B gene: evidence for a direct functional interaction between two classes of regulatory proteins. Genes Dev 6:864–875

Gouil Q, Baulcombe DC (2016) DNA methylation signatures of the plant chromomethyltransferases. PLoS Genet 12:e1006526

Grob S, Schmid MW, Grossniklaus U (2014) Hi-C analysis in Arabidopsis identifies the KNOT, a structure with similarities to the flamenco locus of Drosophila. Mol Cell 55:678–693

Guan JS, Xie H, Ding X (2015) The role of epigenetic regulation in learning and memory. Exp Neurol 268:30–36

Hagmann J, Becker C, Müller J, Stegle O, Meyer RC, Wang G, Schneeberger K, Fitz J, Altmann T, Bergelson J, Borgwardt K, Weigel D (2015) Century-scale methylome stability in a recently diverged *Arabidopsis thaliana* lineage. PLoS Genet 11:e13546

Hauben M, Haesendonckx B, Standaert E, Van Der Kelen K, Azmi A, Akpo H, Van Breusegem F, Guisez Y, Bots M, Lambert B, Laga B, De Block M (2009) Energy use efficiency is characterized by an epigenetic component that can be directed through artificial selection to increase yield. Proc Natl Acad Sci USA 106:20109–920114

Hollick JB (2010) Paramutation and development. Annu Rev Cell Dev Biol 26:557–579

Holliday R (1987) The inheritance of epigenetic defects. Science 238:163–170

Hu L, Li N, Xu C, Zhong S, Lin X, Yang J, Zhou T, Yuliang A, Wu Y, Chen YR, Cao X, Zemach A, Rustgi S, von Wettstein D, Liu B (2014) Mutation of a major CG methylase in rice causes genome-wide hypomethylation, dysregulated genome expression, and seedling lethality. Natl Acad Sci USA 111:10642–10647

Jaligot E, Rival A, Beule T, Dussert S, Verdeil JL (2000) Somaclonal variation in oil palm (*Elaeis guineensis* Jacq): the DNA methylation hypothesis. Plant Cell Rep 19:684–690

Johannes F, Porcher E, Teixeira FK, Saliba-Colombani V, Simon M, Agier N, Bulski A, Albuisson J, Heredia F, Audigier P, Bouchez D, Dillmann C, Guerche P, Hospital F, Colot V (2009) Assessing the impact of transgenerational epigenetic variation on complex traits. PLoS Genet 5:e1000530

Kawakatsu T, Huang SS, Jupe F, Sasaki E, Schmitz RJ, Urich MA, Castanon R, Nery JR, Barragan C, He Y, Chen H, Dubin M, Lee CR, Wang C, Bemm F, Becker C, O'Neil R, O'Malley RC, Quarless DX, 1001 Genomes Consortium, Schork NJ, Weigel D, Nordborg M, Ecker JR (2016) Epigenomic diversity in a global collection of *Arabidopsis thaliana* accessions. Cell 166:492–505

Kinoshita T, Seki M (2014) Epigenetic memory for stress response and adaptation in plants. Plant Cell Physiol 55:1859–1863

Kinoshita Y, Saze H, Kinoshita T, Miura A, Soppe WJ, Koornneef M, Kakutani T (2007) Control of FWA gene silencing in *Arabidopsis thaliana* by SINE-related direct repeats. Plant J 49:38–45

Kooke R, Johannes F, Wardenaar R, Becker F, Etcheverry M, Colot V, Vreugdenhil D, Keurentjes JJ (2015) Epigenetic basis of morphological variation and phenotypic plasticity in *Arabidopsis thaliana*. Plant Cell 27:337–348

Lamke J, Baurle I (2017) Epigenetic and chromatin-based mechanisms in environmental stress adaption and stress memory in plant. Genome Biol 18:124

Law JA, Jacobsen SE (2010) Establishing, maintaining and modifying DNA methylation patterns in plants and animals. Nat Rev Genet 11:204–220

Levenson JM, Sweatt JD (2005) Epigenetic mechanisms in memory formation. Nat Rev Neurosci 6:108–118

Li Q, Eichten SR, Hermanson PJ, Zaunbrecher VM, Song J, Wendt J, Rosenbaum H, Madzima TF, Sloan AE, Huang J, Burgess DL, Richmond TA, McGinnis KM, Meeley RB, Danilevskaya ON, Vaughn MW, Kaeppler SM, Jeddeloh JA, Springer NM (2014) Genetic perturbation of the maize methylome. Plant Cell 26:4602–4616

Li Q, Song J, West PT, Zynda G, Eichten SR, Vaughn MW, Springer NM (2015) Examining the causes and consequences of context-specific differential DNA methylation in maize. Plant Physiol 168:1262–1274

Lindroth AM, Cao X, Jackson JP, Zilberman D, McCallum CM, Henikoff S, Jacobsen SE (2001) Requirement of CHROMOMETHYLASE3 for maintenance of CpXpG methylation. Science 292:2077–2080

Lisch D (2012) Regulation of transposable elements in maize. Curr Opin Plant Biol 15:511–516

Lisch D, Bennetzen JL (2011) Transposable element origins of epigenetic gene regulation. Curr Opin Plant Biol 14:156–161

Lister R, O'Malley RC, Tonti-Filippini J, Gregory BD, Berry CC, Millar AH, Ecker JR (2008) Highly integrated single-base resolution maps of the epigenome in Arabidopsis. Cell 133:523–536

Martinez G, Köhler C (2017) Role of small RNAs in epigenetic reprogramming during plant sexual reproduction. Curr Opin Plant Biol 36:22–28

Matzke MA, Kanno T, Matzke AJ (2015) RNA-directed DNA methylation: the evolution of a complex epigenetic pathway in flowering plants. Annu Rev Plant Biol 66:243–267

Meagher RB (2010) The evolution of epitype. Plant Cell 22:1658–1666

Mirouze M, Reinders J, Bucher E, Nishimura T, Schneeberger K, Ossowski S, Cao J, Weigel D, Paszkowski J, Mathieu O (2009) Selective epigenetic control of retrotransposition in Arabidopsis. Nature 461:427–430

Niederhuth CE, Bewick AJ, Ji L, Alabady MS, Kim KD, Li Q, Rohr NA, Rambani A, Burke JM, Udall JA, Egesi C, Schmutz J, Grimwood J, Jackson SA, Springer NM, Schmitz RJ (2016) Widespread natural variation of DNA methylation within angiosperms. Genome Biol 17:194

Ong-Abdullah M, Ordway JM, Jiang N, Ooi SE, Kok SY, Sarpan N, Azimi N, Hashim AT, Ishak Z, Rosli SK, Malike FA, Bakar NA, Marjuni M, Abdullah N, Yaakub Z, Amiruddin MD, Nookiah R, Singh R, Low ET, Chan KL, Azizi N, Smith SW, Bacher B, Budiman MA, Van Brunt A, Wischmeyer C, Beil M, Hogan M, Lakey N, Lim CC, Arulandoo X, Wong CK, Choo CN, Wong WC, Kwan YY, Alwee SS, Sambanthamurthi R, Martienssen RA (2015) Loss of Karma transposon methylation underlies the mantled somaclonal variant of oil palm. Nature 252:533–537

Ossowski S, Schneeberger K, Lucas-Lledó JI, Warthmann N, Clark RM, Shaw RG, Weigel D, Lynch M (2010) The rate and molecular spectrum of spontaneous mutations in *Arabidopsis thaliana*. Science 327:92–94

Patterson GI, Thorpe CJ, Chandler VL (1993) Paramutation, an allelic interaction, is associated with a stable and heritable reduction of transcription of the maize b regulatory gene. Genetics 135:881–894

Pecinka A, Abdelsamad A, Vu GT (2013) Hidden genetic nature of epigenetic natural variation in plants. Trends Plant Sci 18:625–632

Pikaard CS, Mittelsten Scheid O (2014) Epigenetic regulation in plants. Cold Spring Harb Perspect Biol 6:a019315

Quadrana L, Colot V (2016) Plant transgenerational epigenetics. Annu Rev Genet 50:467–491

Reinders J, Wulff BB, Mirouze M, Marí-Ordóñez A, Dapp M, Rozhon W, Bucher E, Theiler G, Paszkowski J (2009) Compromised stability of DNA methylation and transposon immobilization in mosaic Arabidopsis epigenomes. Genes Dev 23:839–850

Richards EJ (2006) Inherited epigenetic variation – revisiting soft inheritance. Nat Rev Genet 7:395–401

Richardson SR, Morell S, Faulkner GJ (2014) L1 retrotransposons and somatic mosaicism in the brain. Annu Rev Genet 48:1–27

Roux F, Colomé-Tatché M, Edelist C, Wardenaar R, Guerche P, Hospital F, Colot V, Jansen RC, Johannes F (2011) Genome-wide epigenetic perturbation jump-starts patterns of heritable variation found in nature. Genetics 188:1015–1017

Schmitz RJ, Schultz MD, Lewsey MG, O'Malley RC, Urich MA, Libiger O, Schork NJ, Ecker JR (2011) Transgenerational epigenetic instability is a source of novel methylation variants. Science 334:369–373

Secco D, Wang C, Shou H, Schultz MD, Chiarenza S, Nussaume L, Ecker JR, Whelan J, Lister R (2015) Stress induced gene expression drives transient DNA methylation changes at adjacent repetitive elements. Elife 4:e09343. https://doi.org/10.7554/eLife.09343

Seymour DK, Becker C (2017) The causes and consequences of DNA methylome variation in plants. Curr Opin Plant Biol 36:56–63

Seymour DK, Koenig D, Hagmann J, Becker C, Weigel D (2014) Evolution of DNA methylation patterns in the Brassicaceae is driven by differences in genome organization. PLoS Genet 10:e13546

Shen X, De Jonge J, Forsberg SK, Pettersson ME, Sheng Z, Hennig L, Carlborg Ö (2014) Natural CMT2 variation is associated with genome-wide methylation changes and temperature seasonality. PLoS Genet 10:e1004842

Soppe WJ, Jacobsen SE, Alonso-Blanco C, Jackson JP, Kakutani T, Koornneef M, Peeters AJ (2000) The late flowering phenotype of fwa mutants is caused by gain-of-function epigenetic alleles of a homeodomain gene. Mol Cell 6:791–802

Stam M, Belele C, Dorweiler JE, Chandler VL (2002a) Differential chromatin structure within a tandem array 100 kb upstream of the maize b1 locus is associated with paramutation. Genes Dev 16:1906–1918

Stam M, Belele C, Ramakrishna W, Dorweiler JE, Bennetzen JL, Chandler VL (2002b) The regulatory regions required for B' paramutation and expression are located far upstream of the maize b1 transcribed sequences. Genetics 162:917–930

Stelpflug SC, Eichten SR, Hermanson PJ, Springer NM, Kaeppler SM (2014) Consistent and heritable alterations of DNA methylation are induced by tissue culture in maize. Genetics 198:209–218

Stroud H, Ding B, Simon SA, Feng S, Bellizzi M, Pellegrini M, Wang GL, Meyers BC, Jacobsen SE (2013) Plants regenerated from tissue culture contain stable epigenome changes in rice. Elife 2:e00354

Stroud H, Do T, Du J, Zhong X, Feng S, Johnson L, Patel DJ, Jacobsen SE (2014) Non-CG methylation patterns shape the epigenetic landscape in Arabidopsis. Nat Struct Mol Biol 21:64–72

Takuno S, Gaut BS (2012) Body-methylated genes in Arabidopsis thaliana are functionally important and evolve slowly. Mol Biol Evol 29:219–227

Tanurdzic M, Vaughn MW, Jiang H, Lee TJ, Slotkin RK, Sosinski B, Thompson WF, Doerge RW, Martienssen RA (2008) Epigenomic consequences of immortalized plant cell suspension culture. PLoS Biol 6:2880–2895

Verhoeven KJ, Van Dijk PJ, Biere A (2010) Changes in genomic methylation patterns during the formation of triploid asexual dandelion lineages. Mol Ecol 19:315–324

Verhoeven KJF, Macel M, Wolfe LM, Biere A (2011) Population admixture, biological invasions and the balance between local adaptation and inbreeding depression. Proc Biol Sci 278:2–8

Wang C, Liu C, Roqueiro D, Grimm D, Schwab R, Becker C, Lanz C, Weigel D (2015) Genome-wide analysis of local chromatin packing in *Arabidopsis thaliana*. Genome Res 25:246–256

Wibowo A, Becker C, Marconi G, Durr J, Price J, Hagmann J, Papareddy R, Putra H, Kageyama J, Becker J, Weigel D, Gutierrez-Marcos J (2016) Hyperosmotic stress memory in Arabidopsis is mediated by distinct epigenetically labile sites in the genome and is restricted in the male germline by DNA glycosylase activity. Elife 5:e13546

Woo HR, Pontes O, Pikaard CS, Richards EJ (2007) VIM1, a methylcytosine-binding protein required for centromeric heterochromatinization. Genes Dev 12:267–277

Xu J, Tanino KK, Horner KN, Robinson SJ (2016) Quantitative trait variation is revealed in a novel hypomethylated population of woodland strawberry (*Fragaria vesca*). BMC Plant Biol 16:240

Zemach A, Kim MY, Hsieh PH, Coleman-Derr D, Eshed-Williams L, Thao K, Harmer SL, Zilberman D (2013) The Arabidopsis nucleosome remodeler DDM1 allows DNA methyltransferases to access H1-containing heterochromatin. Cell 153:193–205

Zhang X, Yazaki J, Sundaresan A, Cokus S, Chan SW, Chen H, Henderson IR, Shinn P, Pellegrini M, Jacobsen SE, Ecker JR (2006) Genome-wide high-resolution mapping and functional analysis of DNA methylation in Arabidopsis. Cell 126:189–201

Zilberman D (2017) An evolutionary case for functional gene body methylation in plants and animals. Genome Biol 18:87

Zilberman D, Gehring M, Tran RK, Ballinger T, Henikoff S (2006) Genome-wide analysis of Arabidopsis thaliana DNA methylation uncovers an interdependence between methylation and transcription. Nature Genet 39:61–69

Zovkic IB, Guzman-Karlsson MC, Sweatt JD (2013) Epigenetic regulation of memory formation and maintenance. Learn Mem 20:61–74

Plant Accommodation to Their Environment: The Role of Specific Forms of Memory

Michel Thellier, Ulrich Lüttge, Victor Norris, and Camille Ripoll

Abstract A plant germinates and becomes established in a particular place, which remains its permanent location and where it must respond to signals generated by the dynamics of all kinds of external conditions. By putting together old and new data originating from physiology, ecology and epigenetics, it is inferred that the so-called "learning" and "storage/recall" forms of memory are fundamental to the fitness of plants.

1 Introduction

In the course of evolution, plant species have genetically adapted to practically all sorts of soils and climates. However, even when a plant has germinated in a place where it can grow normally, it will have to cope with a variety of stimuli such as wind, rain, touching, length of daylight, cold shocks, heat shocks, drought, wounds inflicted by herbivores, attacks by pests and even electromagnetic radiation (including that emitted by mobile telephones; Tafforeau et al. 2002, 2004; Vian et al. 2006). In most cases, a calcium wave in the cell cytosol is associated with the perception of the stimulus (Knight et al. 1991; Trewavas 1999). The characteristics of the calcium wave (such as its amplitude and duration) are thought to help guide the plant towards an appropriate response (Dolmetsch et al. 1997; McAinch and Hetherington 1998; Knight et al. 1998). The final response may be straightforward, i.e. quasi-immediate

---

M. Thellier (✉)
University of Rouen, Paris, France

U. Lüttge
Department of Biology, Technical University of Darmstadt, Darmstadt, Germany

V. Norris
Faculté des Sciences, Laboratory of Microbiology Signals and Microenvironment, EA 4312, Université de Rouen, Mont-Saint-Aignan Cedex, France

C. Ripoll
Faculté des Sciences, Laboratoire "Assemblages moléculaires, modélisation et imagerie SIMS", Université de Rouen, Mont-Saint-Aignan Cedex, France

© Springer International Publishing AG, part of Springer Nature 2018
F. Baluska et al. (eds.), *Memory and Learning in Plants*, Signaling and Communication in Plants, https://doi.org/10.1007/978-3-319-75596-0_7

and stereotyped; the problem here is that if a range of stimuli were to elicit different responses of the quasi-immediate type, many of these responses might be unnecessary, contradictory and thus energetically costly. Hence, this type of response is best adapted to rare types of stimuli. Memorization of information relative to the perceived stimuli resolves the problem of how to respond to common stimuli that require different, possibly contradictory, responses. More precisely, two specific forms of memory (Trewavas 2003)—the so-called "learning" (also termed "priming" or "training") and "storage/recall" forms of memory—may enable plants to produce a coherent response that is appropriate to the whole range of the stimuli generated by the environment (for data in either of three different European languages, see Thellier 2015, 2017a, b).

2 The "Learning" Form of Plant Memory

In the learning form, the repetition of the same stimulus progressively modifies the intensity of the response, either negatively or positively, and by analogy with animal physiology (Bailey and Chen 1983), such behaviours may be termed "familiarization" (alias "habituation") or "sensitization", respectively. As an example of familiarization, when tobacco plantlets have been subjected to a series of gusts of wind, they transiently cease to produce a cytosolic calcium wave after another gust of wind (Knight et al. 1992). Another example is observed with *Arabidopsis* seedlings, where the cytosolic calcium wave produced by cold shock is attenuated when this shock is preceded by prolonged or repeated cold treatments (Plieth et al. 1999). An example of sensitization, again with *Arabidopsis* seedlings, is that the cytosolic calcium wave produced after a hyperosmotic stress (mimicking drought stress) is increased in plants that have been subjected previously to hyperosmotic stress (Knight et al. 1998). The evolutionary advantage of the learning form of memory is that plants can economize responses to non-dangerous stimuli (in the case of familiarization) and that plants can give stronger responses to dangerous stimuli and stresses (in the case of sensitization).

It was also observed (Knight et al. 1998) that the elevation of cytosolic calcium concentration as a result of hyperosmotic stress in *Arabidopsis* seedlings is reduced by oxidative stress pretreatment, which means that the combination of different sorts of stress can produce novel responses.

3 The "Storage/Recall" Form of Plant Memory

In the storage/recall form of memory (Desbiez et al. 1984), on perception of a stimulus, information is stored within the plant (storage function "on"). However, stored information remains latent (i.e. without any immediate overt effect) until, at a later time, a second, appropriate stimulus enables the plant to recall stored information and let it take effect in controlling its metabolism and development (recall function "on"). The main properties of the storage and recall functions are inferred from data obtained in three independent series of experiments (for reviews, see, e.g. Thellier and Lüttge 2013):

1. When an asymmetrical stimulus (pricking one of the two cotyledons) was administered to *Bidens* seedlings, information was stored as the command that cotyledonary bud growth should be asymmetrical (storage function "on"), namely, that the bud at the axil of the non-pricked cotyledon should start to grow faster than the opposite bud. The cotyledonary bud growth was elicited by removing the terminal bud with a razor blade (plant decapitation). However, the alteration of symmetry of bud outgrowth only occurred if plant decapitation was carried out in the morning (i.e. stored information was expressed, alias recall function "on"), whereas no alteration of symmetry of bud outgrowth occurred when plant decapitation was carried out at midday (i.e. stored information remained unexpressed, alias recall function "off").
2. When a pricking stimulus (whether symmetrical or asymmetrical) was administered to *Bidens* seedlings slightly younger than in case (1), information was stored commanding that hypocotyl elongation should be reduced (storage function "on"). Transferring the seedlings from their nutrient solution to water had the effect of switching "on" the recall function, with the consequence that hypocotyl elongation was then reduced in the pricked plants compared with non-pricked controls. By contrast, hypocotyl elongation was not reduced (compared with non-pricked controls) in the pricked plants when they were left in their nutrient solution (recall function "off").
3. When various non-injurious stimuli were administered to flax seedlings, information was stored as the command to produce epidermal meristems in the hypocotyl (storage function "on"). The recall function was switched "on" when the seedlings were subjected to a temporary calcium shortage, typically for 1 day, whereas a temporary calcium excess prevented meristem production (recall function "off") (Verdus et al. 2012). It is only when the recall function was "on" that meristems were produced in the stimulated seedlings compared with non-stimulated controls.

Clearly, in the series of experiments (1) to (3), the stored information is not the memorization of the stimuli themselves. It is rather an "instruction" or "command" for an appropriate response to these stimuli (for review, see, e.g. Thellier and Lüttge 2013). Plants usually have limited resources; hence, their crucial challenge is how to optimize the allocation of these resources to their principal processes (growth and reproduction, defence, etc.) (Gayler et al. 2008; Matyssek et al. 2012). When a plant is subjected to a stimulus, the instruction stored at first is likely to be a command for modification of resource allocation finally causing the overt reactions as seen in the three types of experiments and in other examples of various plants and stimuli. This may result from a mere locking/unlocking of genes. Indeed, epigenetic control by DNA methylation/demethylation reactions has been found by several independent groups to be associated with the perception of stimuli and stresses (cited in Thellier and Lüttge 2013).

Moreover, in experiments of type (3) with flax seedlings subjected only to a single manipulation stimulus or to a manipulation stimulus followed by three drought stresses, the seedlings responded (after transient calcium shortage) by the production of meristems in greater number in the second than in the first case (Verdus et al. 1997). This experiment is consistent with the idea, which was already apparent in the study of the learning form of memory, that the plant response to a combination of stimuli differs from the response to each of these stimuli separately. The storage function may therefore help the plant generate an integrated response to the entirety of the stimuli that have been perceived over a period of time.

Different stimuli or changes of environmental conditions can enable a plant to recall—or prevent it from recalling—stored information (recall function "on" or "off"). For instance, in experiments of series (3), when flax seedlings have stored the command to form meristems, we have seen that the recall function can be turned "on" or "off" by imposing a transient shortage or excess of calcium, respectively. In the first case, the seedlings will produce meristems; in the second case, they will not. The recall function can thus permit synchronizing the final response of plants to stimulation with the occurrence of other internal or external processes including plant rhythms. In the experiments described above, the turning on/off of the recall function was shown to be linked with ultradian, circadian and annual rhythms (for review, see, e.g. Hütt et al. 2015). Moreover, the recall of stored information may occur repeatedly (at least twice), which enables a plant to produce its response to a given stimulus in different tissues and at different times. This is also shown by the observation that the cotyledon-pricking of *Bidens* seedlings was responsible (1) for the inhibition of hypocotyl elongation after plant transfer to water (recall function "on") when the seedlings were a few days old and (2) for specifying, a few days later, the precedence between the cotyledonary buds depending on the time of the day when the terminal bud was removed.

Discussion Most experiments as described above have been made on young, nonwoody plants under laboratory conditions. The question then arises: are the above data valid only under these restrictive conditions or do they have a broader significance? A few studies carried out under conditions closer to the natural ones may be on the way.

An example of learning is given by the leaf-folding behaviour of the sensitive plant *Mimosa pudica* in response to the mechanical signals from trampling megaherbivores; this behaviour is a defensive reflex that allows the plant to avoid being seen. There is a trade-off between protection and productivity because in the folded state, photosynthesis is drastically reduced. This is particularly detrimental at low and limiting photosynthetically active radiation (PAR) but less so at high PAR. The response of plants maintained under low- and high-PAR environments showed different patterns of learning. The low-PAR plants learned faster to ignore the stimulus and retained the memory of this longer, i.e. for up to a month when undisturbed, than the high-PAR plants. Thus, in the trade-off, the PAR-limited plants choose photosynthesis rather than defence, whilst the high-PAR plants choose defence because a reduction in photosynthesis is less disadvantageous (Gagliano et al. 2014).

As regards the storage/recall form of memory, long-term storage of information has been shown to exist in bryony (*Bryonia dioica*), a common cucurbitaceous species which is climbing in bushes and hedges. When the terminal internode of a bryony plant is gently rubbed during its elongation, it becomes shorter and thicker, and its equipment in peroxidase enzymes is increased compared with non-rubbed control internodes (Boyer et al. 1979). With tissue cultures obtained from fragments of internodes of bryony, increased peroxidase activities were found to exist in several successive subcultures derived from stimulated internodes (Bourgeade et al. 1989): the information induced by the rubbing stimulus was thus stored during at least several months. Another example of long-term storage comes from the woody plant (*Rosa hybrida*). An appropriate stimulus inhibited the growth of buds formed after the stimulus but did not inhibit the growth of pre-existing buds (Grémiaux et al. 2016); a simple interpretation of this observation is that an instruction to change the allocation of resources (so as to inhibit bud growth) was stored on perception of the stimulus and that this instruction was recalled only on initiation of formation of a new bud.

4 Conclusions

Plants and animals are both endowed with memories, though these memories have different functions and operate via different mechanisms and substrates. For animals, which are mobile, the strategy for optimizing their chances to remain alive and reproduce consists in exploring their surroundings in order to find environmental conditions that suit them best. They need to be able to memorize a huge quantity of data (places, things, time lapses, feelings, etc.) to orient themselves in space and time when they wander around. This memory depends on the fact that they possess neurons with tremendous capacities and (for the highest species) a central nervous system with its billions of interconnected neurons.

For plants, which cannot move around to find optimal environmental conditions, the strategy consists in adjusting their metabolism and development to the conditions existing at the place where they have rooted. Having no neurons, the way plants memorize information—and the extent to which they do this—must differ from that of animals. Indeed, the mere interplay of genes may suffice to explain how learning and the storage/recall capacities of memory allow plants to adjust the intensity of their response to the dangerousness of a stimulus, to produce an integrated response to the variety of stimuli perceived along time and/or to synchronize their response to a stimulus with the occurrence of other external and internal events including rhythms. However rudimentary plant memories may be compared to animal memories, they have thus the potential to play a crucial part in the accommodation of plants to their environment.

References

Bailey C, Chen M (1983) Morphological basis of long-term habituation and sensitization in *Aplysia*. Science 220:91–93

Bourgeade P, Boyer N, De Jaegher G, Gaspar T (1989) Carry-over of thigmomorphogenetic characteristics in calli derived from *Bryonia dioica* internodes. Plant Cell Tissue Organ Cult 19:199–211

Boyer N, Gaspar T, Lamand M (1979) Modification des isoperoxydases et de l'allongement des entre-noeuds de bryone à la suite d'irritations mécaniques. Z Pflanzenphysiol 93:459–470

Desbiez MO, Kergosien Y, Champagnat P, Thellier M (1984) Memorization and delayed expression of regulatory messages in plants. Planta 160:392–399

Dolmetsch RE, Lewis RS, Goodnow CC, Healy JJ (1997) Differential activation of transcription factors induced by Ca^{2+} response amplitude and duration. Nature 386:855–858

Gagliano M, Renton M, Depczynski M, Mancuso S (2014) Experience teaches plants to learn faster and forget slower in environments where it matters. Oecologia 175:63–72

Gayler S, Grams TEE, Heller W, Treutter D, Priesack E (2008) A dynamic model of environmental effects on allocation to carbon-based secondary compounds in juvenile trees. Ann Bot 101:1089–1098

Grémiaux A, Girard S, Guérin V, Lothier J, Baluska F, Davies E, Bonnet P, Vian A (2016) Low-amplitude, high frequency electromagnetic field exposure causes delayed and reduced growth in *Rosa hybrid*a. J Plant Physiol 190:44–53

Hütt MT, Lüttge U, Thellier M (2015) Noise-induced phenomena and complex rhythms: a test scenario for plant systems biology. In: Mancuso S, Shabala S (eds) Rhythms in Plants. Springer, Switzerland, pp 279–321

Knight MR, Campbell AK, Smith SM, Trewavas AJ (1991) Transgenic plant aequorin reports the effect of touch and cold-shock and elicitors on cytoplasmic calcium. Nature 352:524–526

Knight MR, Smith SM, Trewavas AJ (1992) Wind-induced plant motion immediately increases cytosolic calcium. Proc Natl Acad Sci USA 89:4967–4971

Knight H, Brandt S, Knight MR (1998) A history of stress alters drought calcium signalling pathway in Arabidopsis. Plant J 16:681–687

Matyssek R, Koricheva J, Schnyder H, Ernst D, Munch JC, Osswald W, Pretzsch H (2012) The balance between resource sequestration and retention: a challenge in plant science. In: Matyssek R, Schnyder H, Osswald W, Ernst D, Munch JC, Pretzsch H (eds) Growth and defence in plants – Resource allocation at multiple scales, Ecological Studies. Springer, Berlin, pp 3–24

McAinch MR, Hetherington AM (1998) Encoding specificity in Ca^{2+} signalling systems. Trends Plant Sci 3:32–36

Plieth C, Hansen UP, Knight H, Knight MR (1999) Temperature sensing by plants. The primary characteristics of signal perception and calcium response. Plant J 18:491–497

Tafforeau M, Verdus MC, Norris V, White G, Demarty M, Thellier M, Ripoll C (2002) SIMS study of the calcium-deprivation step related to epidermal meristem production induced in flax by cold shock or radiation from a GSM telephone. J Trace Microprobe Tech 20:611–623

Tafforeau M, Verdus MC, Norris V, White GJ, Cole M, Demarty M, Thellier M, Ripoll C (2004) Plant sensitivity to low intensity 105 GHz electromagnetic radiation. Bioelectromagnetics 25:403–407

Thellier M (2015) Les plantes ont elles une mémoire ? Quæ. Versailles, France

Thellier M (2017a) Plant responses to environmental stimuli, the role of specific forms of plant memory. Quæ, Versailles, France and Springer, Dordrecht, The Netherlands

Thellier M (2017b) Haben Pflanzen ein Gedächtnis? Quæ, Versailles, France and Springer, Heidelberg, Germany [Translation from Thellier 2015 by Ulrich Lüttge]

Thellier M, Lüttge U (2013) Plant memory: a tentative model. Plant Biol 15:1–12

Trewavas A (1999) *Le calcium c'est la vie*: calcium waves. Plant Physiol 120:1–6

Trewavas A (2003) Aspects of plant intelligence. Ann Bot 92:1–20

Verdus MC, Thellier M, Ripoll C (1997) Storage of environmental signals in flax. Their morphogenetic effect as enabled by a transient depletion of calcium. Plant J 12:1399–1410

Verdus MC, Ripoll C, Norris V, Thellier M (2012) The role of calcium in the recall of stored morphogenetic information by plants. Acta Biotheoretica 60:83–97

Vian A, Roux D, Girard S, Bonnet P, Paladian F, Davies E, Ledoigt V (2006) Microwave irradiation affects gene expression in plants. Plant Signal Behav 1:67–70

Memristors and Electrical Memory in Plants

Alexander G. Volkov

Abstract Plants have different forms of memory, such as sensory, short, and long-term memory. Memory is the ability to store the state of a system at a given time and access this information at a later time. Possible candidates for memory in plants are memristors, which are resistors with memory. Voltage-gated K^+ channels have memristive properties in seeds, plants, flowers, and fruits. The Venus flytrap, which has memristors, can accumulate small subthreshold charges, and when the threshold value is reached, the trap closes. The cumulative character of electrical stimuli points to the existence of short-term electrical memory in the Venus flytrap. As soon as the 8 µC charge for a small trap or a 9 µC charge for a large trap is transmitted between a lobe and midrib from the external capacitor, the trap starts to close at room temperature. At temperatures 28–36 °C, a smaller electrical charge of 4.1 µC is required to close the trap of the *Dionaea muscipula*. Sensory and long-term memory were also found in the Venus flytrap. Plants have biological clocks and circadian rhythms, which involve electrical elements of memory. Many plants are able to memorize daytime and nighttime. The biological clock in plants is sensitive to light, which can reset the phase of the rhythm.

1 Plant Memory

The general classification of memory in plants is based on the duration of memory retention and identifies three distinct types of memory: sensory memory, short-term memory, and long-term memory (Trewavas 2003, 2005, 2014). Sensory memory corresponds approximately to the initial 0.1–3.0 s after a signal is perceived. Some information in sensory memory can be transferred then to short-term memory. Short-term memory allows one to recall something from several seconds to as long as a minute without rehearsal. The storage in sensory memory and short-term memory generally has a strictly limited capacity and duration, which means information is available for a certain period of time but is not retained indefinitely. Long-term

A. G. Volkov (✉)
Department of Chemistry, Oakwood University, Huntsville, AL, USA

memory can store much larger quantities of information for potentially unlimited duration up to a whole life span of the plant.

Many examples of memory and learning have been observed in plants, including storage and recall functions in seedlings (Thellier et al. 2000; Thellier and Lüttge 2013), chromatin remodeling in plant development (Amasino 2004; Goodrich and Tweedie 2002; Reyes et al. 2002; Sung and Amasino 2004), transgeneration memory of stress (Bruce et al. 2007; Goh et al. 2003; Molinier et al. 2006), immunological memory of tobacco plants (Baldwin and Schmelz 1996; Conrath 2006) and mountain birches (Ruuhola et al. 2007), vernalization and epigenetic memory of winter (Amasino 2004; Sung and Amasino 2004, 2006), induced resistance and susceptibility to herbivores (Karban 2008; Karban and Niiho 1995; Ruuhola et al. 2007), memory response in ABA-entrained plants (Goh et al. 2003), phototropically and gravitotropically induced memory in maize (Nick et al. 1990; Nick and Schafer 1988), ozone sensitivity of grapevine as a memory effect in a perennial crop plant (Soja et al. 1997), memory of mechanical stimulus (Ueda et al. 2007; Ueda and Nakamura 2006; Volkov et al. 2007, 2008a, b; Gagliano et al. 2014), systemic acquired resistance in plants exposed to a pathogen (Conrath 2006), and electrical memory (Markin and Volkov 2012; Volkov et al. 2007, 2008a, b, c, d). Sensory and short-term memories in the Venus flytrap are caused by electrical phenomena (Volkov et al. 2008c, 2009a, b, 2011a).

2 Venus Flytrap: Electrical Memory

Electrical signaling, memory, rapid trap closure, digestion, and slow trap opening of the carnivorous plant *Dionaea muscipula* Ellis (Venus flytrap) have been attracting the attention of researchers since the nineteenth century (Darwin 1875; Burdon-Sanderson 1873, 1874; Brown 1916; Brown and Sharp 1910; Lloyd 1942; Munk 1876; Volkov 2012a, b). When an insect touches the piezoelectric hairs (Fig. 1), these mechanosensors trigger a receptor potential, which generate an electrical signal that acts as an action potential (Volkov et al. 2007). Two stimuli generate two action potentials, which activate the trap closing at room temperature in a fraction of a second. At high temperatures of 28–36 °C, only one mechanical stimulus is required for the trap closing. Propagation of action potentials and the trap closing can be blocked by inhibitors of voltage-gated channels (Volkov et al. 2008d). The two mechanical stimuli required for the trap closing should be applied within an interval from 0.75 s to 40 s (Volkov et al. 2008a, c, 2009a, b).

Electrical signaling and memory play fundamental roles in plant responses. The Venus flytrap can accumulate small subthreshold charges, and when the threshold value is reached, the trap closes. The cumulative character of electrical stimuli points to the existence of short-term electrical memory in the Venus flytrap (Volkov et al. 2008c, 2009a, 2011a, b). Using charge injection method (Volkov et al. 2007, 2008a, b, c), it was evident that the application of an electrical stimulus between the midrib (positive potential) and a lobe (negative potential) causes the Venus flytrap to close

Fig. 1 Mechanosensitive trigger hairs in the Venus flytrap

the trap without any mechanical stimulation. The average stimulation pulse voltage sufficient for rapid closure of the Venus flytrap was 1.5 V. Probably, electrostimulation activated ion pumps and/or voltage-gated channels in the midrib.

The inverted polarity pulse with negative voltage applied to the midrib did not close the plant. Applying impulses in the same voltage range with inverted polarity did not open the trap, even with pulses of up to 100 s. As soon as the 8 µC charge for a small trap or a 9 µC charge for a large trap is transmitted between a lobe and midrib from the external capacitor, the trap starts to close at room temperature (Fig. 2a). At temperatures 28–36 °C, a smaller electrical charge of 4.1 µC is required to close the trap of the *Dionaea muscipula* (Fig. 2b). Probably, there is a phase transition at temperature between 25 °C and 28 °C, and due to this reason, only one mechanical stimulus or a small electrical charge is required to close the trap. The cumulative character of electrical stimuli points to the existence of short-term electrical memory in the Venus flytrap (Markin et al. 2008; Volkov et al. 2009a,b).

It was found that energy for trap closure is generated by ATP hydrolysis (Jaffe 1973) and used for a fast transport of cations (Rea 1983). The amount of ATP drops from 950 µM per midrib before mechanical stimulation to 650 µM per midrib after stimulation and closure (Jaffe 1973). However, it is not clear if electrical stimulation triggers the closing process or contributes energy to the closing action. The action potential delivers the electrical signal to the midrib, which can activate the trap closing (Fig. 3).

Brown (1916) indicated that strong electrical shock between lower and upper leaves can cause the Venus flytrap to close, but in their article, the amplitude and polarity of applied voltage, charge, and electrical current were not reported. The trap did not close when we applied the same electrostimulation between the upper and lower leaves as we applied between a midrib and a lobe, even when the injected

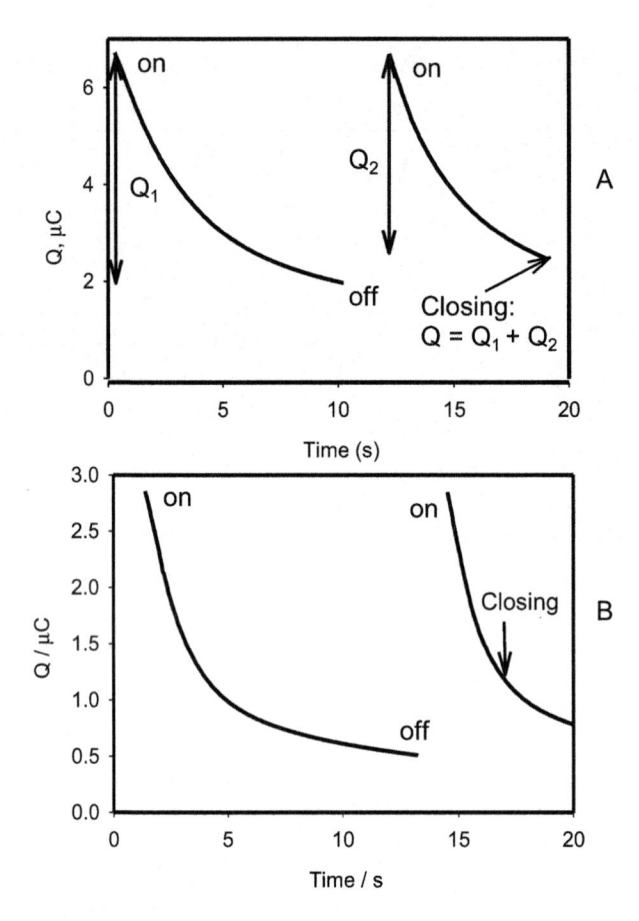

Fig. 2 Closing of the *Dionaea muscipula* trap at 20 °C (**a**) and 32 °C (**b**) by electrical charge Q injected between a lobe and a midrib. Electrical discharge was measured using NI-PXI-4071 digital multimeter

charge was increased from 14 μC to 750 μC. It is probable that the electroshock induced by Brown (1916) had a very high voltage and electrical current.

Sensory memory was found in the Venus flytrap (Volkov et al. 2009a, b). Insects can touch a few trigger hairs or just one for 2 s. In biology, this effect is referred to as molecular or sensory memory. It was shown that sustained membrane depolarization induced a "molecular" memory phenomenon and has profound implication on the biophysical properties of voltage-gated ion channels. One sustained mechanical stimulus applied to only one trigger hair can close the trap of *Dionaea muscipula* (Volkov et al. 2009a) in a few seconds (Fig. 4). Prolonged pressing of the trigger hair generates two electrical signals like action potentials within an interval of about 2 s, which stimulate the trap closing (Fig. 4).

Memristors and Electrical Memory in Plants 143

Fig. 3 Chain of events in the trap closing and opening

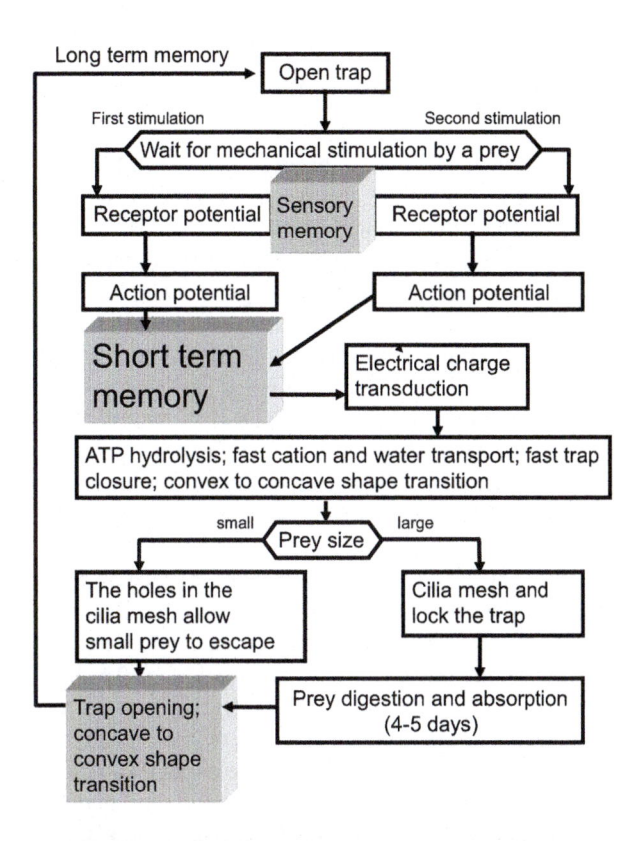

Fig. 4 Closing of the *Dionaea muscipula* trap by pressing of only one trigger hair at 20 °C

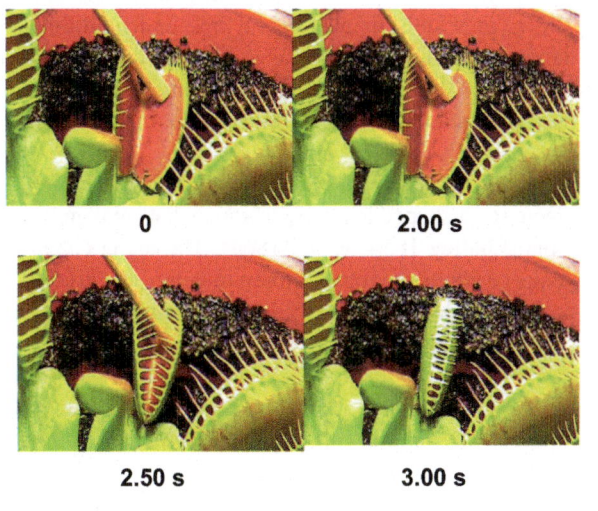

3 Circadian Rhythms in Biologically Closed Electrical Circuits of Plants: Time Sensing and Memory

The circadian clock was discovered by De Mairan (1729) in his first attempt to resolve experimentally the origin of rhythm in the leaf movements of *Mimosa pudica*. This rhythm continued even when *Mimosa pudica* was maintained under continuous darkness. De Mairan (1729) hypothesized that the *Mimosa pudica* leaf movement is controlled by a biological clock.

In plants, circadian rhythms are linked to the light-dark cycle. Many of the circadian rhythmic responses to day and night continue in constant light or dark, at least for a period of time. The circadian clock is an endogenous oscillator with a period of approximately 24 h. The circadian clock in plants is sensitive to light, which can reset the phase of the rhythm (Volkov et al. 2011a, c). Molecular mechanism underlying circadian clock function is poorly understood, although it is now widely accepted for both plants and animals that it is based on circadian oscillators.

Wang et al. (2012) found that the molecular circadian clock and redox states regulate the electrical activity of potassium channels in suprachiasmatic nucleus neurons in mammalian central circadian clock. Circadian rhythms in plants also have electrical components (Volkov et al. 2011a, c). The biological clock regulates a wide range of electrophysiological and developmental processes in plants. There is the direct influence of a circadian clock on biologically closed electrochemical circuits in vivo (Volkov et al. 2011a, c). The biologically closed electrochemical circuits in the leaves of *C. miniata* (Kaffir lily), *Aloe vera*, and *Mimosa pudica*, which regulate their physiology, were analyzed using the charge stimulation method (Volkov et al. 2011a, c). Plants memorize daytime and nighttime. Even at continuous light or darkness, plants recognize nighttime or daytime and change the input resistance (Volkov et al. 2011a, c). The circadian clock can be maintained endogenously and has electrochemical oscillators, which can activate ion channels in biologically closed electrochemical circuits.

A monocotyledon *Clivia miniata* (Lindl.) Regel is a vascular plant with dark green, strap-shaped leaves with somewhat swollen leaf bases which arise from fleshy roots (Lindley 1854; Regel 1864). The name *Clivia* was given to this ornamental genus of plants by John Lindley to compliment Lady Clive, the Duchess of Northumberland, who first cultivated this plant in England. *C. miniata* is a model for the study of circadian rhythms in plants.

Electrical responses of *C. miniata* to electrical stimulation were measured during the day in daylight, darkness at night, and the following day in darkness with different timing and voltages using the charge stimulation method (Volkov et al. 2011c).

As it is well known, if a capacitor of capacitance C with initial voltage V_0 is discharged through a resistor R, the voltage at time t is

$$V(t) = V_0 \cdot e^{-t/\tau} \qquad (1)$$

where

$$\tau = RC \qquad (2)$$

denotes the time constant. Equation (1) in logarithmic form reads

$$\ln(V(t)/V_0) = -t/t \qquad (3)$$

The time constant, τ, can be determined from the slope of this linear function. The circuit time constant τ governs the discharging process. As the capacitance or resistance increases, the time of the capacitor discharge increases according to equation (1). The resistance of the linear circuit can be easily found from Eq. (3).

However if the function (3) is not linear, then one can find the so-called input resistance at any moment of the time (Volkov et al. 2010c, 2011c):

$$R_{\text{input}} = -\frac{U}{C \; dU/dt} \qquad (4)$$

The electrostimulation of the leaves of *C. miniata* (Kaffir lily) was provided with different voltages and electrical charges (Volkov et al. 2011c). Resistance between Ag/AgCl electrodes in the leaf of *C. miniata* was higher at night than during the day or the following day in the darkness. The biologically closed electrical circuits with voltage-gated ion channels in *C. miniata* are activated the next day, even in the darkness. *C. miniata* memorizes daytime and nighttime. At continuous light, *C. miniata* recognizes nighttime and increases the input resistance to the nighttime value even under light. These results show that the circadian clock can be maintained endogenously and has electrochemical oscillators, which can activate voltage-gated ion channels in biologically closed electrochemical circuits. The activation of voltage-gated channels depends on the applied voltage, electrical charge, and speed of transmission of electrical energy from the electrostimulator to the *C. miniata* leaves (Volkov et al. 2011a, c).

The difference in kinetics of a capacitor discharge between night and day can be also demonstrated by subtracting corresponding responses one from another. Figure 5a shows the difference "night minus day" in kinetics of a capacitor discharge. The difference between night and the following day in the darkness is presented in Fig. 5b. These differences are very similar proving that the biological clock in *Clivia* recognizes the daytime, even in darkness. During the day when the lights are on, the results are exactly the same, which were reproduced on different *Clivia* plants.

The *Mimosa pudica* Linn. is a nyctinastic plant that closes its leaves in the evening. The process occurs when the pinnules fold together and the whole leaf droops downward temporarily until sunrise. The leaves open in the morning due to a circadian rhythm, which is regulated by a biological clock with a cycle of about 24 h.

Fig. 5 (a) Difference in time dependencies of the capacitor discharge in the *C. miniata* leaf at night and at daytime (response during the day subtracted from the response during the night). (b) Difference in time dependencies of electrical discharge in the *C. miniata* leaf at night and in the dark during daytime (response in the dark during daytime subtracted from the response during the night). V is the capacitor voltage and V_0 is the initial voltage in volts. Polarity and location of electrodes along the leaf are shown

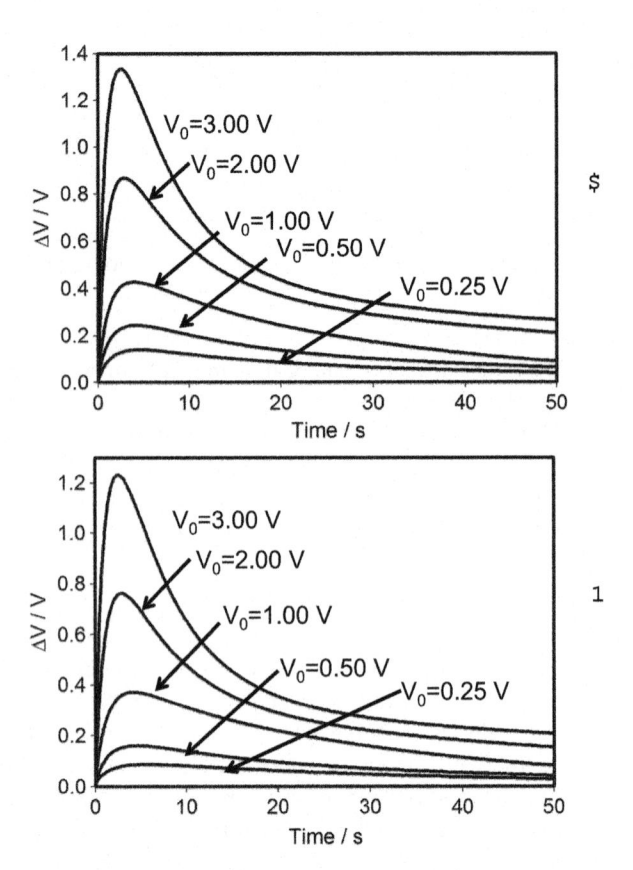

During photonastic movement in the *Mimosa pudica*, leaves recover their daytime position. During a scotonastic period, the primary pulvini straighten up, and pairs of pinnules fold together about the tertiary pulvini. The closing of pinnae depends upon the presence of phytochrome in the far-red absorbing form (Fondeville et al. 1966).

Leaf movement in the *Mimosa pudica* appears to be regulated by electrical signal transduction. Mechanics of these movements are hidden in the specialized organ— the pulvinus. The pulvinus is a thickened organ at the base of the leaf or leaflet, which is a motor organ for leaf movement.

Isolated pulvinar protoplasts are responsive to light signals in vitro (Coté 1995; Kim et al. 1992, 1993). In the dark period, the closed inward-directed K^+ channels of extensor cells are opened within 3 min by blue light. Conversely, the inward-directed K^+ channels of flexor cells, which are open in the darkness, are closed by blue light. In the light period, however, the situation is more complex. Premature darkness alone is sufficient to close the open channels of extensor protoplasts, but both darkness and a preceding pulse of red light are required to open the closed channels in the flexor protoplasts (Kim et al. 1992, 1993).

Recently, we analyzed kinetics of a capacitor discharge during a day in *Mimosa pudica* and found the input resistance between electrodes inserted across the

Fig. 6 Difference in time dependencies of the capacitor discharge (1 μF, 1.0 V) in a pulvinus of *Mimosa pudica* at night and daytime (night–day response)

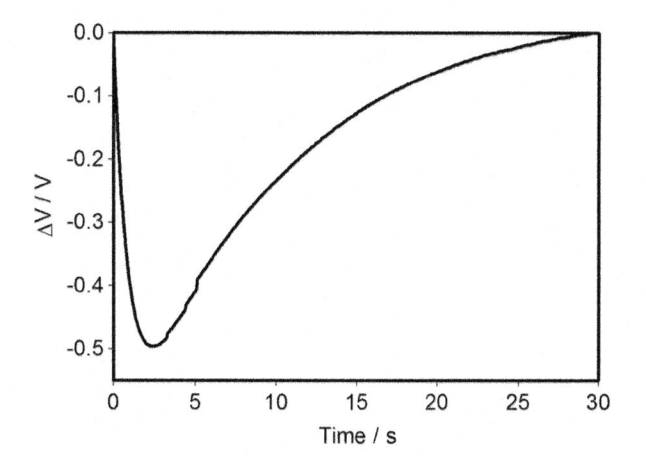

pulvinus (Volkov et al. 2010a, b, c, d, 2011a). We also investigated the electrical activity of *Mimosa pudica* in the daylight, *at night*, and in the darkness the following day. There is difference in the discharge kinetics between day and nighttime (Fig. 6). The initial difference in the speed of the response (faster during the day) can be explained by activation of ion channels, equivalent to the high rectification effect. This effect depends on the applied stimulation voltage. The kinetics of a capacitor discharge depends on the polarity of electrodes indicating the electrical rectification in the *Mimosa pudica*. This dependence can be explained by a change in resistivity with applied potential due to opening of voltage-gated ion channel. Opening of voltage-gated channels induces the effect of electrical rectification which was found in *Aloe vera* (Volkov et al. 2010c), the Venus flytrap (Volkov et al. 2009b), and *Mimosa pudica* (Volkov et al. 2010b) during the day. Volkov et al. (2010b) used a silicon rectifier Schottky diode NTE583 as a model of a voltage-gated ion channel and reproduced experimental dependencies of a capacitor discharge in plant tissue of *Mimosa pudica*.

Figure 7 shows the difference in time dependence of a capacitor discharge in a pulvinus of *Mimosa pudica* during the night and the daytime but in the dark. *Mimosa pudica* was exposed to a 12:12 h light/dark photoperiod during a 2-month period, but in the morning of the experiment, the light source was not switched on for the day. The biologically closed electrical circuits with voltage-gated ion channels in *Mimosa pudica* are activated the next day even in the darkness. This phenomenon can be caused by biological clock in *Mimosa pudica*. The nonlinear effect of activation of electrical circuits during the daytime is stronger than during the next day in the darkness.

Circadian oscillators are components of the biological clocks that regulate the activities of plants in relation to environmental cycles and provide an internal temporal framework. The circadian clock regulates a wide range of electrophysiological and developmental processes in plants. Plant tissues have biologically closed electrochemical circuits that are involved in these regulations. We found periodic

Fig. 7 Difference in time dependencies of a capacitor electrical discharge (1 μF, 1.0 V) in a pulvinus of *Mimosa pudica* at night and in the dark at daytime (night-dark day response)

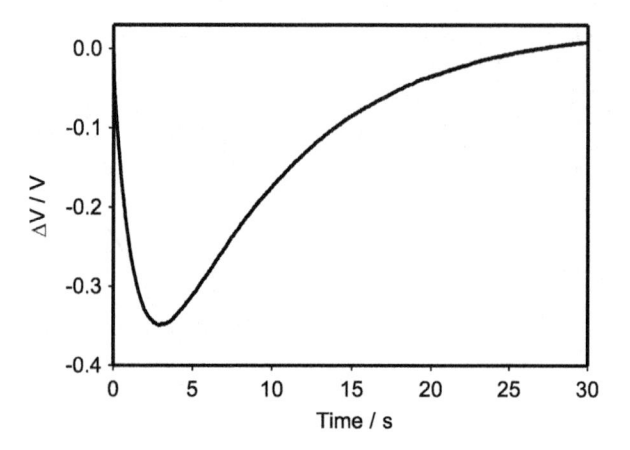

activation and deactivation of these circuits in *Mimosa pudica*, *C. miniata* (Kaffir lily), and *Aloe vera* controlled by internal clock, rather than environmental clues.

This circadian rhythm can be related to the difference in the membrane potentials during the day and nighttime, which was found in pulvini of different plants (Kim et al. 1992, 1993; Kumon and Tsurumi 1984; Racusen and Satter 1975; Scott and Gulline 1975). During the day in darkness, there is still a rectification effect in the pulvinus; however, in the presence of light during the daytime, this rectification effect and the resistance decrease in the pulvinus are two times higher. These results demonstrate that the circadian clock can be maintained endogenously, probably involving electrochemical oscillators, which can activate or deactivate ion channels in biologically closed electrochemical circuits.

4 Electrical Memory Elements: Memristors

Memristors are memory circuit elements whose properties depend on the history and state of the system (Chua 1971). Memristors can participate in electrical signal transduction along plasma membranes between phytosensors and phytoactuators (Volkov 2017). One of such memristors can be a voltage-gated K^+ channel according to a pharmaceutical analysis (Volkov et al. 2014a, b, c, d). Memristors can be involved in plant electrical memory. Sensory and short-term memories in the Venus flytrap have electrical components.

Plant membranes have a specific property, called excitability. This property is used by cells, tissues, and organs to change their internal condition and external reactions under the action of various environmental factors, referred to as irritants. Electrical processes play important roles in the electrophysiology of plants. Electrical signals propagate along sophisticated electrical circuitry of plants consisting of many electrical components developed by nature. Plasma membranes in plants and axons have resistors, capacitors, memristors, and batteries (Volkov 2017).

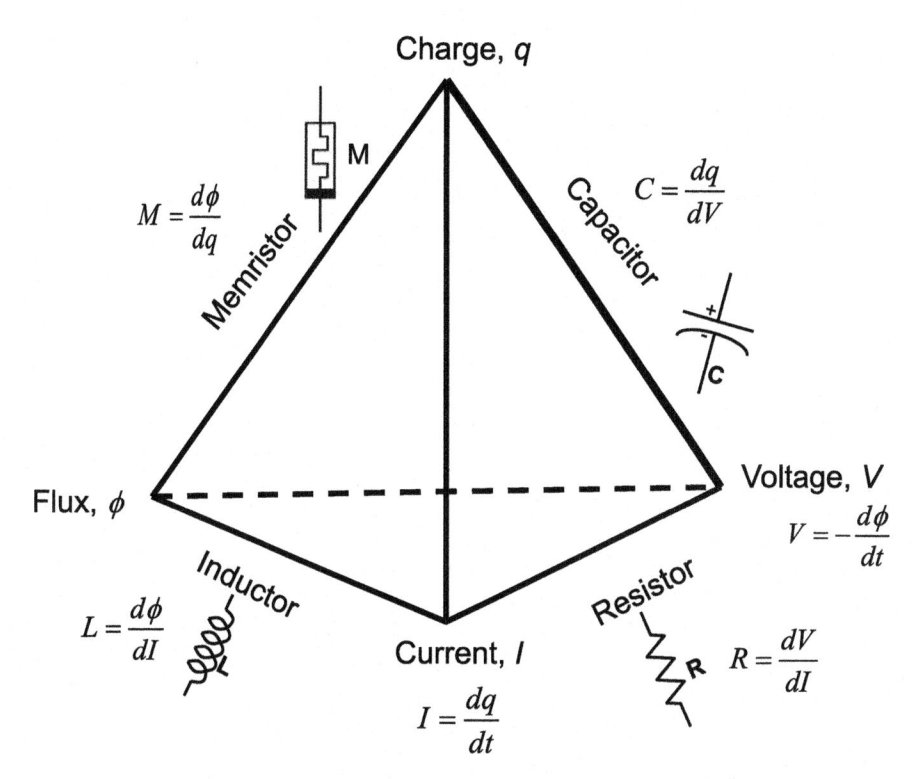

Fig. 8 Four basic electrical circuit elements

The standard electrical circuits comprise of four basic elements: a capacitor, a resistor, an inductor, and a memristor (Fig. 8). The fourth basic circuit element is a memristor, or a resistor with memory, which was discovered by Leon Chua (1971). A memristor is a nanoscale memory device, which has huge potential technical applications in computer science and electronics.

A memristor is a nonlinear element; its current-voltage characteristic is similar to that of a Lissajous pattern (Chua 1971, 2014). No combination of nonlinear resistors, capacitors, and inductors can reproduce this Lissajous behavior of the memristor. It is a fundamental two-terminal electrical circuit element described by the state-dependent Ohm's law (Chua 1971). A voltage-controlled memristor can be defined by

$$I = G(x_1, x_2 \ldots x_n; V)V$$
$$\frac{dx_k}{dt} = f_k(x_1, x_2 \ldots x_n; V), k = 1, 2, \ldots n \quad (5)$$

where G is the memductance of the memristor. The state variables $(x_1, x_2, \ldots, x_n)$ depend on the internal state of the memristor and are defined by "n" first-order differential equations called the associated state equations. A current-controlled memristor is defined by

$$V = M(x_1, x_2 \ldots x_n; I)I$$
$$\frac{dx_k}{dt} = f_k(x_1, x_2 \ldots x_n; I), k = 1, 2, \ldots n \qquad (6)$$

where M is the memristance of the memristor. The unit of the memristance is the ohm. The unit of the memductance is the Siemens (Chua 1971, 2014).

Mathematically memristance can be described by the equation

$$M(q(t)) = \frac{d\varphi(q)}{dq} = \frac{\left(\frac{d\varphi}{dt}\right)}{\frac{dq}{dt} = \frac{V(t)}{I(t)}} \qquad (7)$$

where φ and q denote the flux and charge, respectively (Chua 2014).

The power of a memristor is

$$P(t) = I^2(t)M(q(t)) = I(t)V(t), \qquad (8)$$

During the last decade, different memristors were developed as semiconductor devices, enzymatic systems, polymers, and electrified interfaces (Hota et al. 2012; Johnsen et al. 2011; MacVittie and Katz 2013; Pershin et al. 2009). Theoretical analysis shows the existence of memristors in neural networks, voltage-gated channels, synapses, and the brain. Chua et al. (2012), Chua (2014), and Sah et al. (2013) presented theoretical proofs that the voltage-gated K^+ channel is a locally active memristor in neurons.

Since plants and animals have similar voltage-gated K^+ channels, one can expect the possible presence of memristors in plants. Recently, memristors were found as components of plasma membranes in many plants, flowers, fruits, and seeds (Volkov 2017; Volkov et al. 2014a, b, c, d, 2015; Volkov et al. 2016a, b; Volkov and Markin 2016; Volkov and Nyasani 2016). The analysis of the presence of memristors in a bio-tissue is based on cyclic voltammetric characteristics where the memristor, a resistor with memory, should manifest itself. Tetraethylammonium chloride, an inhibitor of voltage-gated K^+ channels, or 5-nitro-2-(3-phenylpropylamino)benzoic acid (NPPB), a blocker of voltage-gated Cl^- and K^+ channels (Illek et al. 1993), transforms a memristor to a resistor in plant tissue (Volkov 2017; Volkov et al. 2014a, b, c, d, 2015, Volkov et al. 2016a, b; Volkov and Markin 2016; Volkov and Nyasani 2016). Uncouplers carbonyl cyanide-3-chlorophenylhydrazone (CCCP) and carbonyl cyanide-4-trifluoromethoxy-phenylhydrazone (FCCP) decrease the amplitude of electrical responses at low and high frequencies of bipolar periodic electrostimulating waves. The discovery of memristors in the plant kingdom creates a new direction in the modeling and understanding of electrical phenomena in plant membrane structures.

Markin et al. (2014) created a mathematical model of memristors with capacitors in plants. Adhikaru et al. (2013) found that a memristor has three characteristic fingerprints: "When driven by a bipolar periodic signal the device must exhibit a pinched hysteresis loop in the voltage-current plane, assuming the response is periodic; starting from some critical frequency, the hysteresis lobe area should decrease monotonically as the excitation frequency increases; the pinched hysteresis loop should shrink to a single-valued function when the frequency tends to infinity." The pinched hysteresis loop transforms to a non-pinched hysteresis loop in membrane systems instead of a single line $I = V/R$ at high frequencies of the applied voltage because the amplitude of electrical current depends also on capacitance, frequency, and direction of scanning:

$$I = C\frac{dV}{dt} \qquad (9)$$

The pinched hysteresis loop of memory elements, when subject to a periodic stimulus, can be self-crossing (memristor type 1) or not (memristor type 2). Volkov et al. (2014a, b, c, d) found that the electrostimulation of plants by bipolar periodic sinusoidal or triangular waves induces electrical responses with fingerprints of memristors of type 1 or type 2. The analytical model of biological generic memristors developed by Markin et al. (2014) predicts fingerprints of a generic memristor connected in parallel to a capacitor and resistor:

1. The pinched hysteresis loop should shrink to a single-valued function if frequency of a bipolar periodic signal is very low.
2. The device must exhibit a pinched hysteresis loop in the voltage-current plane when driven by a bipolar periodic signal at low frequency.
3. The pinched hysteresis disappears according to Equation (9) if frequency of a bipolar periodic signal is very high.

The third fingerprint of a generic memristor can be explained by an obvious fact that the biological membranes with ion channels had capacitance connected in parallel to transmembrane organized ion channels. Such type of memristing devices with a capacitance is called a *generic* memristor.

Hodgkin and Huxley (1952) formulated a membrane model that accounts for K^+, Na^+, and ion leakage channels in the squid giant axon. The membrane resting potential for each ion species is treated like a battery, and the degree to which the channel is open is modeled by a variable resistor. According to Chua, the Hodgkin-Huxley time-varying potassium conductance is in fact a first-order memristor, and the Hodgkin-Huxley time-varying sodium conductance is in fact a second-order memristor. Nayak and Sikdar (2007) found time-dependent molecular memory in single voltage-gated sodium channel. Volkov (2017), Volkov et al. (2014a, b, c, d, 2015, 2016a, b), and Volkov and Markin (2016) found that voltage-gated K^+ channels in plants, seeds, fruits, and flowers are memristors.

5 Memristors in Plasma Membrane of Plants, Fruits, Seeds, and Flowers

Memristors are memory circuit elements whose properties depend on the history and state of the system. Memristors can participate in electrical signal transduction between phytosensors and phytoactuators (Volkov 2017).

We investigated the possible presence of memristors in the electrical circuitry of the different plants and fruits using experimental setup shown in Fig. 9. Electrical current through plants or fruits was estimated according to Ohm's law:

$$I = V_R/R \tag{10}$$

The electrostimulation by bipolar sinusoidal or triangle periodic waves induces electrical responses in potato tubers, *Mimosa pudica*, *Aloe vera*, apple fruits, flowers, and the Venus flytrap with fingerprints of memristors (Volkov Volkov 2017; Volkov et al. 2014a, b, c, d, 2015, Volkov et al. 2016a, b; Volkov and Markin 2016). The same results were obtained using electrochemical potentiostat. The analysis was based on cyclic voltammetric characteristics where the memristor should manifest itself.

Memory is the ability to store the state of a system at a given time and access this information at a later time. Candidates for electrical memory in plants are memristors and memcapacitors, which are resistors and capacitors with memory.

Electrochemical processes play important roles in the bioelectrochemistry of fruits, seeds, roots, and plants. Electrical form of energy can be used to do biochemical and mechanical work or in information transfer and analysis. These signals

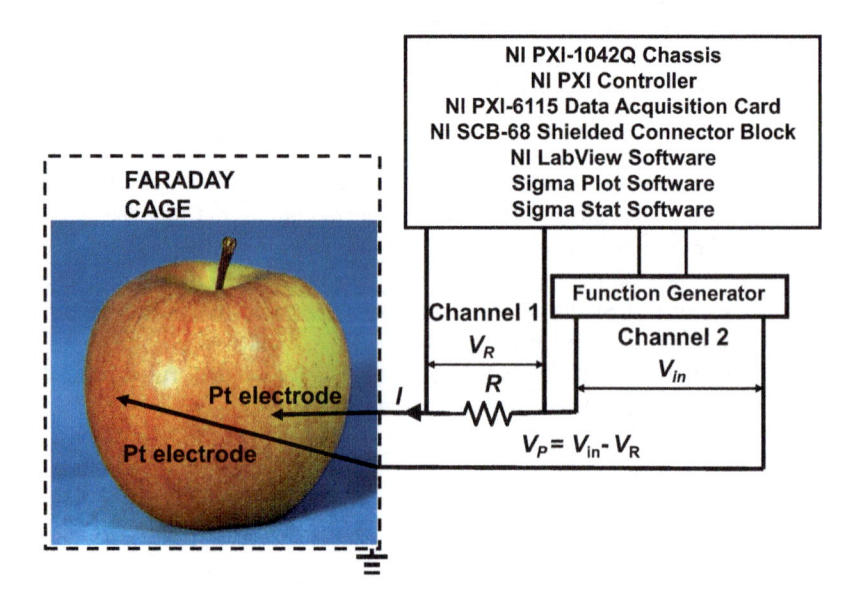

Fig. 9 Schematic diagram of the data acquisition and electrostimulation system

propagate along sophisticated electrical circuitry of plants consisting of many electrical components developed by nature (Baluška et al. 2006; Ksenzhek and Volkov 1998). A potential pathway for transmission of electrical signals on long distances might be the phloem sieve tube system since it represents a continuum of plasma membranes (Volkov 2000).

Apple fruits are the most common and well-studied fruits. Apple fruit (*Malus domestica* Borkh.) is one of the important sink organs in sugar accumulation. Zhang et al. (2004) presented results on anatomy of an apple and ultrastructure of conductive bundles and phloem cells. Lang and Ryan (1994) analyzed phloem and xylem tissue in apple fruits. Velasco et al. (2010) decoded the complete genome of the Golden Delicious apple and found about 57,000 genes. Kurenda et al. (2013) studied effects of injection of ion channel inhibitors such as TEACl, gadolinium chloride, 9-anthracenecarboxylic acid, and diethylstilbestrol on the inhibition of bio-speckle activity in apple tissue. The electrical current flowing through the Gala fruit, generated by a bipolar triangular wave with amplitude of ±4 V, is shown in Fig. 10. Figure 10 presents a pinched hysteresis loop in the voltage-current plane. The plot displays a common pinched point with self-crossing between curves with coordinates $I = 0$ µA and $V_P = 0$ V (Fig. 10). There is a pinched hysteresis loop in the voltage-current plane with self-crossing between curves with coordinates $I = 4$ µA and $V_P = 2.3$ V (Fig. 10).

Increasing the sweep rate of a bipolar triangular wave changes the shape of the line: it is still a loop but without a pinched point. The electrostimulation of the Gala apple fruit by a periodic bipolar triangular wave induces electrical responses with fingerprints of a memristor. Similar results were obtained for electrostimulation of a Golden Delicious and Arkansas Black apple fruits by a bipolar sinusoidal or triangular wave (Volkov et al. 2016b).

Electrostimulation of the apple fruit by a bipolar sinusoidal wave also induces a pinched hysteresis loop in the voltage-current plane at low frequencies of electrostimulation. The plot displays a common pinched point with self-crossing between curves with coordinates $I = 0$ µA and $V_P = 0$ V. There is a pinched hysteresis loop in the voltage-current plane with self-crossing between curves with coordinates $I = 3$ µA and $V_P = 2.3$ V.

Electrical networks in plants and fruits consist of different electrical circuits. Electrical signals can propagate to adjacent excitable cells due to the electrical coupling between them, which is the major path for cell-to-cell electrical coupling. Voltage-gated ion channels can regulate generation and transduction of electrical signals along membranes. It was intriguing to investigate if blockers of ionic channels can change characteristics or even the presence of memristors in fruits.

Figure 11 shows the effect of tetraethylammonium chloride (TEACl) on electrical responses induced by periodic bipolar sinusoidal wave from a function generator with platinum electrodes inserted along vascular bundles in an apple fruit. A 2 mL of 0.3 M TEACl was injected in the Gala apple 70 h before electrical measurements. A 70-h delay was required because of the slow distribution and partition of TEACl in the apple tissue. TEACl starts to work after the injection to the apple tissue, but for the maximum effect of a complete inhibition, we need to wait for partition of an

Fig. 10 Cyclic voltammetry of a Gala apple fruit. Frequency of bipolar triangular voltage wave was 0.1 mHz with the sweep rate of 1.8 mV/s (**a**) and 1 Hz with the sweep rate of 18 V/s (**b**). The simple equivalent electrical circuit is shown

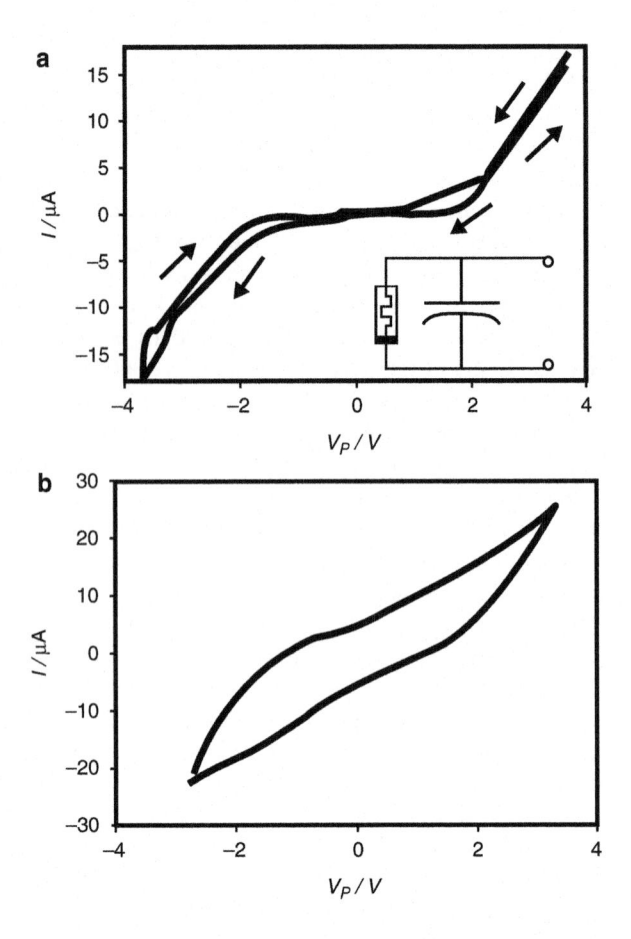

inhibitor in the apple tissue, which is controlled by diffusion. TEACl decreases the amplitude of electrical current between electrodes in the apple tissue, and the pinched hysteresis loop disappears (Fig. 11). Tetraethylammonium chloride, an inhibitor of voltage-gated K^+ channels, transforms a memristor to a conventional resistor in plant tissue at low and high frequencies of bipolar periodic sinusoidal or triangular waves. Passive membrane permeability for K^+ ions varies from 10^{-13} to 10^{-15} cm/s (Volkov et al. 1998). Ion channels can transport a few millions of ions per second. Inhibition of voltage-gated K^+ ion channels in apple fruits increases resistance and decreases electrical current between Pt electrodes in apple tissue (Fig. 11). Injection of 2 mL of distilled water in control experiments does not inhibit memristive properties of apples. Apple fruits have the voltage-gated ion channels associated with plasma membranes. A blocker of the voltage-gated potassium ion channels, TEACl inhibits the memristive properties of the Gala apple fruits (Fig. 11). It means that the voltage-gated potassium ion channels in these apples can be memristors. However, TEACl is not specific and could block some nonselective cation channels.

Fig. 11 Cyclic voltammetry of a Gala apple fruit. Frequency of bipolar triangular voltage scanning was 1 mHz. A 2.5 mL of 0.3 M TEACl was injected by a syringe in an apple 70 h before measurements

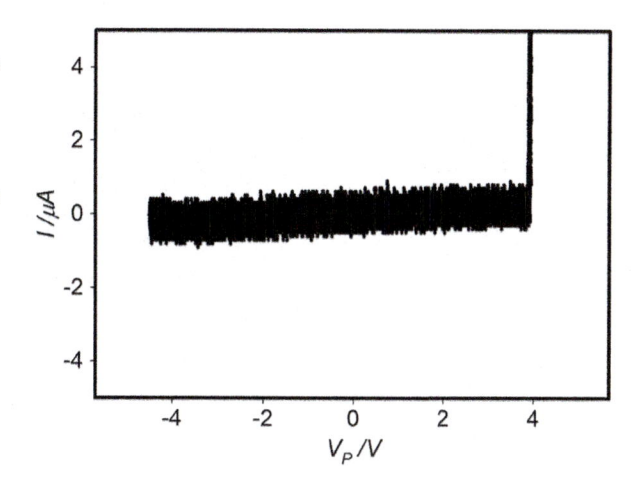

Fig. 12 Electric current I versus voltage V_P applied between a lobe and a midrib of the Venus flytrap, induced by a bipolar sinusoidal voltage wave from a function generator. Frequency of voltage scanning was 0.5 mHz

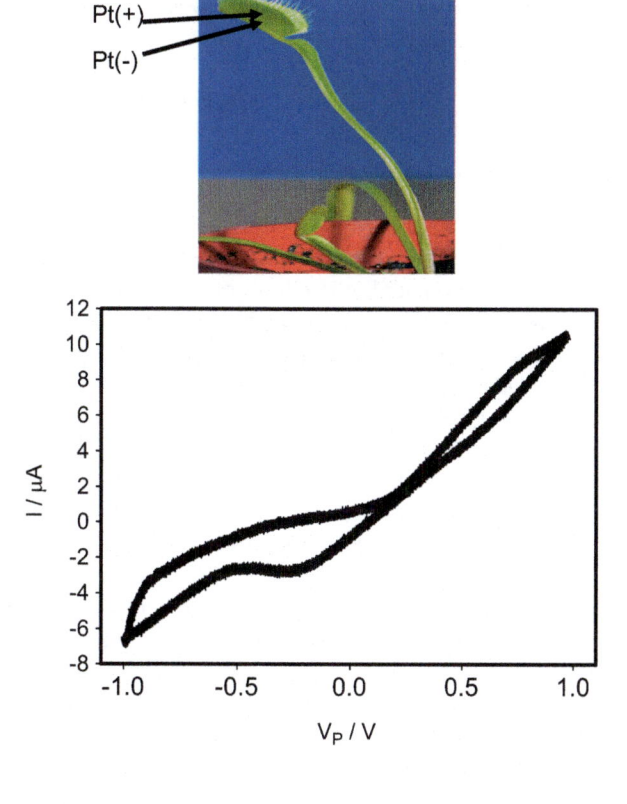

Volkov et al. (2014a, b, c, d) recorded the current flowing through the plant generated by a bipolar sinusoidal wave with frequency of 0.5 mHz for an open trap (Fig. 12) and for a closed trap. In both cases a pinched hysteresis loop was obtained in the voltage-

Fig. 13 Electric current I versus voltage V_P applied along a pulvinus of *Mimosa pudica*, induced by a bipolar sinusoidal voltage wave from a function generator. Frequency of sinusoidal voltage was 0.5 mHz

current plane with one important difference. If the trap is closed with a Pt reference electrode in the midrib, the plot displays a common pinched point with self-crossing between curves, which corresponds to properties of a memristor type 1. If the trap is closed with a Pt reference electrode in the lobe, the plot displays a common pinched point without self-crossing between curves, which corresponds to properties of a memristor type 2. If the trap is open with a Pt reference electrode in the midrib, the plot also displays a common pinched point but without self-crossing between curves, which corresponds to properties of a memristor type 2 (Fig. 12). Increasing of a bipolar sinusoidal wave frequency to 1 kHz changes the shape of the line: it is still a loop but without a pinched point for both open and closed traps. The branches of the loop are approximately parallel to each other. The electrostimulation of the Venus flytrap by a periodic wave induces electrical responses in the Venus flytrap with fingerprints of a memristor of a type 1 (closed trap) and type 2 (open trap, Fig. 12).

Cyclic voltammetry in Fig. 13 shows how electrical current depends on voltage induced by bipolar sinusoidal wave at different frequencies when platinum electrodes are inserted across the pulvinus of the *Mimosa pudica*. At low frequencies of scanning, the plot displays a common pinched point with self-crossing between curves (Fig. 13). The common pinched point disappears at high frequencies of 1 kHz, and memristor transforms to a resistor.

Figure 14 shows how electrical current depends on the voltage V_p induced by bipolar sinusoidal wave from a function generator with a frequency of 0.5 mHz,

Fig. 14 Electric current I versus voltage V_P applied to a leaf of *Aloe vera*, induced by a bipolar sinusoidal voltage wave from a function generator. Frequency of voltage scanning was 0.5 mHz

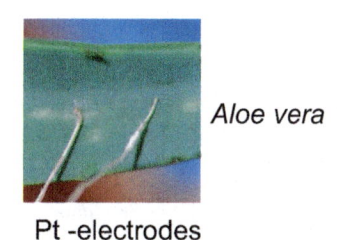

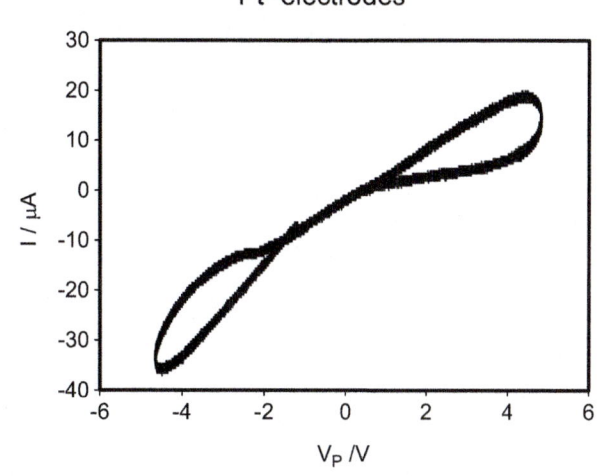

when platinum electrodes are inserted along the vascular bundles in a leaf of *Aloe vera*. There is a self-crossing between curves and a pinched point in hysteresis loop at low frequency of sinusoidal wave in the voltage-current plane when $I = 0$ μA and $V_P = 0$ V, which is a typical sign of a memristor of a first kind (Fig. 14). The increasing of a sinusoidal wave frequency to 100 Hz leads to the disappearing of a pinched point in the complete agreement with characteristics of memristors, which had a small "parasitic" capacitor connected across the memristor. This capacitance can be a function of membrane, electrodes, and plant tissue capacitances. We found that a self-crossing between curves exists in 71% of experiments.

The electrophysiology of plants should include memristors as essential model building blocks in electrical networks. The memristor is an "ideal" circuit element, and in plant tissue memristors coexist with membrane capacitors. The discovery of memristors in plant membranes creates a new direction in the modeling and understanding of electrical phenomena in plant kingdom. It can be a starting point for understanding mechanisms of memory, learning, circadian rhythms, and biological clocks.

Acknowledgment Author would like to thank the Taylor & Francis Publishing House. This chapter is derived in part from our articles (Volkov et al. 2014a, b, c, d, 2015; Volkov and Nyasani 2016) published in Plant Signaling and Behavior, copyright Taylor & Francis, available online.

References

Adhikaru AP, Sah MP, Kim H, Chua L (2013) Three fingerprints of memristor. IEEE Trans Circuits Syst 60:3008–3021
Amasino R (2004) Vernalization, competence, and the epigenic memory of winter. Plant Cell 16:2553–2559
Baldwin IT, Schmelz EA (1996) Immunological "memory" in the induced accumulation of nicotine in wild tobacco. Ecology 77:236–246
Baluška F, Mancuso S, Volkman D (eds) (2006) Communication in plants. Neuronal aspects of plant life. Springer, Berlin
Brown WH (1916) The mechanism of movement and the duration of the effect of stimulation in the leaves of *Dionaea*. Am J Bot 3:68–90
Brown WH, Sharp LW (1910) The closing response in Dionaea. Bot Gaz 9:290–302
Bruce TJA, Matthes MC, Napier J, Pickett JA (2007) Stressful "memories" of plants: evidence and possible mechanisms. Plant Sci 173:603–608
Burdon-Sanderson J (1873) Note on the electrical phenomena which accompany stimulation of the leaf of *Dionaea muscipula* Ellis. Philos Proc R Soc Lond 21:495–496
Burdon-Sanderson J (1874) Venus fly-trap (*Dionaea muscipula*). Nature 10:105–107
Chua L (1971) Memristor—the missing circuit element. IEEE Trans Circuit Theory 18:507–519. https://doi.org/10.1109/TCT.1971.1083337
Chua L (2014) If it's pinched it's a memristor. Semicond Sci Technol 29:104001-1-42
Chua L, Sbitnev V, Kim H (2012) Hodgkin-Huxlew axon is made of memristors. Int J Bifurcat Chaos 22:1230011-1-48. https://doi.org/10.1142/S0218127412500988
Conrath U (2006) Systemic acquired resistance. Plant Signal Behav 1:179–184
Coté GG (1995) Signal transduction in leaf movement. Plant Physiol 109:729–734
Darwin C (1875) Insectivorous plants. Murray, London
De Mairan M (1729) Observation botanique. Histoire de l'Academie Royale de Sciences, Paris, pp 35–36
Fondeville JC, Bortwick HA, Hendrichs SB (1966) Leaflet movement of *Mimosa pudica* L. Indicative of phytochrome action. Planta 69:357–364
Gagliano M, Renton M, Depczynski M, Mancuso S (2014) Experience teaches plants to learn faster and forget slower in environments where it matters. Oecologia 175:63–72
Goh CH, Nam HG, Park YS (2003) Stress memory in plants: a negative regulation of stomatal response and transient induction of rd22 gene to light in abscisic acid-entrained Arabidopsis plants. Plant J 36:240–255
Goodrich J, Tweedie S (2002) Remembrance of things past: chromatin remodeling in plant development. Annu Rev Cell Dev Biol 18:707–746
Hodgkin AL, Huxley AF (1952) A quantitative description of membrane current and its application to conduction and excitation in nerve. J Physiol 117:500–544
Hota MK, Bera MK, Kundu B, Kundu SC (2012) A natural silk fibroin protein-based transparent bio-memristor. Adv Funct Mater 22:4493–4499. https://doi.org/10.1002/asfm.2012200073
Illek B, Fischer H, Kreusel KM, Hegel U, Claus W (1993) Volume-sensitive basolateral K^+ channels in HT-29/B6 cells: block by lidocaine, quinidine, NPPB, and Ba^{2+}. Am J Phys 263: C674–C683
Jaffe MJ (1973) The role of ATP in mechanically stimulated rapid closure of the Venus's flytrap. Plant Physiol 51:17–18
Johnsen GK, Lutken CA, Martinsen OG, Grimnes S (2011) Memristive model of electro-osmosis in skin. Phys Rev E 83:031916. https://doi.org/10.1103/PhysRevE.83.031916
Karban R (2008) Plant behaviour and communication. Ecol Lett 11:1–13
Karban R, Niiho C (1995) Induced resistance and susceptibility to herbivory: plant memory and altered plant development. Ecology 76:1220–1225

Kim HY, Coté GG, Crain RC (1992) Effect of light on the membrane potential of protoplasts from *Samanea saman* pulvini. Involvement of K^+ channels and the H^+-ATPase. Plant Physiol 99:1532–1539

Kim HY, Coté GG, Crain RC (1993) Potassium channels in *Samanea saman* protoplasts controlled by phytochrome and the biological clock. Science 260:960–962

Ksenzhek OS, Volkov AG (1998) Plant energetics. Academic, San Diego

Kumon K, Tsurumi S (1984) Ion efflux from pulvinar cells during slow downward movements of the petiole of *Mimosa pudica* L. induced by photostimulation. J Plant Physiol 115:439–443

Kurenda A, Pieczywek PM, Adamiak A, Zdunek A (2013) Effect of cytochalasin B, lantrunculin B, colchine, cyclohexinid, dimethyl sulfoxide and ion channel inhibitors on biospeckle activity in apple tissue. Food Biophys 8:290–296

Lang A, Ryan KG (1994) Vascular development and sap flow in apple pedicels. Ann Bot 74:381–388

Lindley J (1854) New plant Vallota miniata. Gardener's Chron 8:119

Lloyd FE (1942) The carnivorous plants. Ronald, New York

MacVittie K, Katz E (2013) Electrochemical systems with memimpedance properties. J Phys Chem C 117:24943–24947. https://doi.org/10.1021/jp409257v

Markin VS, Volkov AG (2012) Morphing structures in the Venus flytrap. In: Volkov AG (ed) Plant electrophysiology—signaling and responses. Springer, Berlin, pp 1–31

Markin VS, Volkov AG, Jovanov E (2008) Active movements in plants: mechanism of trap closure by *Dionaea muscipula* Ellis. Plant Signal Behav 3:778–783

Markin VS, Volkov AG, Chua L (2014) An analytical model of memristors in plants. Plant Signal Behav 9:e972887

Molinier J, Ries G, Zipfel C, Hohn B (2006) Transgeneration memory of stress in plants. Nature 442:1046–1049

Munk H (1876) Die electrischen und Bewegungserscheinungen am Blatte der *Dionaeae muscipula*. Arch Anat Physiol Wiss Med:30–203

Nayak TK, Sikdar SK (2007) Time-dependent molecular memory in single voltage-gated sodium channel. J Membr Biol 219:19–36

Nick P, Schafer E (1988) Spatial memory during the tropism of maize (*Zea mays* L.) coleoptiles. Planta 175:380–388

Nick P, Sailer K, Schafer E (1990) On the relation between photo- and gravitropically induced spatial memory in maize coleoptiles. Planta 181:385–392

Pershin YV, La Fontaine S, Di Ventra M (2009) Memristive model of amoeba learning. Phys Rev E 80:021926.0. https://doi.org/10.1103/PhysRevE.80.021926

Racusen R, Satter RL (1975) Rhytmic and phytochrome-regulated changes in transmembrane potential in *Samanea* pulvini. Nature 255:408–410

Rea PA (1983) The dynamics of H^+ efflux from the trap lobes of *Dionaea muscipula* Ellis (Venus's flytrap). Plant Cell Environ 6:125–134

Regel E (1864) *Clivia miniata* Lindl. Amaryllideae. Gartenflora 14:131–134

Reyes JC, Hennig L, Gruissem W (2002) Chromatin-remodeling and memory factors. New regulators of plant development. Plant Physiol 130:1090–1101

Ruuhola T, Salminen JP, Haviola S, Yang S, Rantala MJ (2007) Immunological memory of mountain birches: effects of phenolics on performance of the autumnal moth depend on herbivory history of trees. J Chem Ecol 33:1160–1176

Sah M, Kim H, Chua L (2013) Brains are made of memristors. IEEE Circuits Syst Mag 14:12–36

Scott BIH, Gulline HF (1975) Membrane changes in a circadian system. Nature 254:69–70

Soja G, Eid M, Gangl H, Redl H (1997) Ozone sensitivity of grapevine (*Vitis vinifera* L.): evidence for a memory effect in a perennial crop plant? Phyton 37:265–270

Sung S, Amasino RM (2004) Vernalisation and epigenetics: how plants remember winter. Curr Opin Plant Biol 7:4–10

Sung S, Amasino RM (2006) Molecular genetic study of the memory of winter. J Exp Bot 57:3369–3377

Thellier M, Lüttge U (2013) Plant memory: a tentative model. Plant Biol 15:1–12
Thellier M, Sceller LL, Norris V, Verdus MC, Ripoll C (2000) Long-distance transport, storage and recall of morphogenetic information in plants. The existence of a sort of primitive plant "memory". C R Acad Sci Paris 323:81–91
Trewavas A (2003) Aspects of plant intelligence. Ann Bot 92:1–20
Trewavas A (2005) Green plants as intelligent organisms. Trends Plant Sci 10:414–419
Trewavas A (2014) Plant behaviour and intelligence. Oxford University Press, Oxford
Ueda M, Nakamura Y (2006) Metabolites involved in plant movement and "memory": nyctinasy of legumes and trap movement in the Venus flytrap. Nat Prod Rep 23:548–557
Ueda M, Nakamura Y, Okada M (2007) Endogenous factors involved in the regulation of movement and "memory" in plants. Pure Appl Chem 79:519–527
Velasco R, Zharkikh A, Affourtit J, Dhingra A, Cestaro A, Kalyanaraman A, Fontana P, Bhatnagar SK, Troggio M, Pruss D et al (2010) The genome of the domesticated apple (*Malus x domestica* Borkh). Nat Genet 42:833–839
Volkov AG (2000) Green plants: electrochemical interfaces. J Electroanal Chem 483:150–156
Volkov AG (2012a) Plant electrophysiology: methods and cell electrophysiology. Springer, Berlin
Volkov AG (2012b) Plant electrophysiology: signaling and responses. Springer, Berlin
Volkov AG (2017) Biosensors, memristors and actuators in electrical networks of plants. Int J Parallel Emergent Distrib Syst 32:44–55. https://doi.org/10.1080/17445760.2016.1141209
Volkov AG, Markin VS (2016) Memristors in biomembranes. In: Iglic A, Kulkarni CV, Rappolt M (eds) Advances in biomembranes and lipid self-assembly, vol 24. Academic Press/ABL, UK, pp 91–117
Volkov AG, Nyasani EK (2016) Sunpatiens compact hot coral: memristors in flowers. Funct Plant Biol 45(2):222–227. https://doi.org/10.1071/FP16326
Volkov AG, Deamer DW, Tanelian DI, Markin VS (1998) Liquid interfaces in chemistry and biology. Wiley, New York
Volkov AG, Adesina T, Jovanov E (2007) Closing of Venus flytrap by electrical stimulation of motor cells. Plant Signal Behav 2:139–144
Volkov AG, Adesina T, Jovanov E (2008a) Charge induced closing of *Dionaea muscipula* Ellis trap. Bioelectrochemistry 74:16–21
Volkov AG, Adesina T, Markin VS, Jovanov E (2008b) Kinetics and mechanism of *Dionaea muscipula* trap closing. Plant Physiol 146:694–702
Volkov AG, Carrell H, Adesina T, Markin VS, Jovanov E (2008c) Plant electrical memory. Plant Signal Behav 3:490–492
Volkov AG, Coopwood KJ, Markin VS (2008d) Inhibition of the *Dionaea muscipula* Ellis trap closure by ion and water channels blockers and uncouplers. Plant Sci 175:642–649
Volkov AG, Carrell H, Baldwin A, Markin VS (2009a) Electrical memory in Venus flytrap. Bioelectrochemistry 75:142–147
Volkov AG, Carrell H, Markin VS (2009b) Biologically closed electrical circuits in Venus flytrap. Plant Physiol 149:1661–1667
Volkov AG, Foster JC, Ashby TA, Walker RK, Johnson JA, Markin VS (2010a) *Mimosa pudica*: electrical and mechanical stimulation of plant movements. Plant Cell Environ 33:163–173
Volkov AG, Foster JC, Markin VS (2010b) Signal transduction in *Mimosa pudica*: biologically closed electrical circuits. Plant Cell Environ 33:816–827
Volkov AG, Foster JC, Markin VS (2010c) Molecular electronics in pinnae of *Mimosa pudica*. Plant Signal Behav 5:826–831
Volkov AG, Foster JC, Baker KD, Markin VS (2010d) Mechanical and electrical anisotropy in *Mimosa pudica* pulvini. Plant Signal Behav 5:1211–1221
Volkov AG, Baker K, Foster JC, Clemmons J, Jovanov E, Markin VS (2011a) Circadian variations in biologically closed electrochemical circuits in *Aloe vera* and *Mimosa pudica*. Bioelectrochemistry 81:39–45

Volkov AG, Pinnock MR, Lowe DC, Gay MS, Markin VS (2011b) Complete hunting cycle of *Dionaea muscipula*: consecutive steps and their electrical properties. J Plant Physiol 168:109–120

Volkov AG, Wooten JD, Waite AJ, Brown CR, Markin VS (2011c) Circadian rhythms in electrical circuits of *Clivia miniata*. J Plant Physiol 168:1753–1760

Volkov AG, Forde-Tuckett V, Reedus J, Mitchell CM, Volkova MI, Markin VS, Chua L (2014a) Memristor in the Venus flytrap. Plant Signal Behav 9:e29204

Volkov AG, Reedus J, Mitchell CM, Tuckett C, Volkova MI, Markin VS, Chua L (2014b) Memristor in the electrical network of *Aloe vera* L. Plant Signal Behav 9:e29056

Volkov AG, Reedus J, Mitchell CM, Tuckett C, Volkova MI, Markin VS, Chua L (2014c) Memory elements in the electrical network of *Mimosa pudica* L. Plant Signal Behav 9:e982029

Volkov AG, Tuckett C, Reedus J, Volkova MI, Markin VS, Chua L (2014d) Memristors in plants. Plant Signal Behav 9:e28152-1-8

Volkov AG, Nyasani EK, Blockmon AL, Volkova MI (2015) Memristors: memory elements in potato tubers. Plant Signal Behav 10:e1071750

Volkov AG, Nyasani EK, Tuckett C, Blockmon AL, Reedus J, Volkova MI (2016a) Cyclic voltammetry of apple fruits: memristors *in vivo*. Bioelectrochemistry 112:9–15

Volkov AG, Nyasani EK, Tuckett C, Greeman EA, Markin VS (2016b) Electrophysiology of pumpkin seeds: memristors in vivo. Plant Signal Behav 11:e115600

Wang TA, Yu YV, Govindaiah G, Ye X, Artinian L, Coleman TP, Sweedler JV (2012) Circadian rhythm of redox state regulates excitability in suprachiasmatic nucleus neurons. Science 337:839–842

Zhang LY, Peng YB, Pelleschi-Travier S, Fan Y, Lu YF, Lu YM, Gao XP, Shen YY, Delrot S, Zhang DP (2004) Evidence for apoplasmic phloem unloading in developing apple fruit. Plant Physiol 135:574–586

Towards Systemic View for Plant Learning: Ecophysiological Perspective

Gustavo M. Souza, Gabriel R. A. Toledo, and Gustavo F. R. Saraiva

> *The world is richer than it is possible to express in any single language.*
>
> Ilya Prigogine

Abstract Herein, we have proposed a concept of plant learning based on some principles of systemic plant ecophysiology. In order to accomplish this task, a framework consisting in basic epistemological assumptions is offered, as well as a cognitive context that underpins the perspective of learning. Accordingly, a number of empirical studies are quoted to illustrate the basic idea presented herein.

1 Introduction

In what "world" does it make sense to attribute to plants characteristics common to humans and animals, such as intelligence, learning, decision-making, etc.? This rhetorical question can be written in a more orthodox mode: In what scientific paradigm, in Thomas Kuhn's sense, can concepts related to cognition (learning, attention, memory, decision-making, etc.) be adequately attributed to plants?

This question is critical because science, as a human enterprise seeking for the knowledge and understanding of the universe, is not deprived of a human view of the world. Despite the relative objectivity of scientific knowledge, the philosophical perspective of the observer, even unconsciously, has a powerful influence on the whole process of scientific production, from the questions being asked to the interpretation of the observations (empirical data). Therefore, the question if plants

---

G. M. Souza (✉) · G. R. A. Toledo
Department of Botany, Federal University of Pelotas, Pelotas, Brazil
e-mail: gmsouza@ufpel.br.edu

G. F. R. Saraiva
Research Center for Ecophysiology of West of São Paulo, São Paulo, Brazil

are cognitive entities capable to learn, make decisions and exhibit intelligence, somehow, depends on the scientific perspective of the observer. Obviously, such possibility of understanding on the plant realm must be based on facts, observations of the physical aspects of the plants world, supported by solid scientific theories. Notwithstanding, in the process of the development of new perspectives of the world, analogical thinking and even a little bit of metaphorical inspiration are acceptable and even necessary (Trewavas 2007). For the philosopher of science Paul Feyerabend (1975), the proliferation of theories is beneficial to science, while uniformity weakens its critical power. Accordingly, the condition of coherence, by virtue of which new hypotheses are necessary to conform to accepted theories, is irrational because it preserves the older, not necessarily the better theory.

In this context, plant neurobiology was presented as a proposal of a new science concerning with a more holistic perspective of the plant realm, imputing to them some cognitive attributes (Trewavas 2003; Brenner et al. 2006; Baluška and Mancuso 2007; Barlow 2008; Calvo 2016). Despite the welcome criticism from part of plant biology scientific community (Firn 2004; Alpi et al. 2007; Struik et al. 2008), new ideas and hypothesis are a fundamental condition for the development of science.

Herein, we will focus particularly on the aspects of plant learning. However, to accomplish this task properly, we will make explicit the epistemological framework underlying our approach, in order to specify about what plant model (theoretical representation) we are talking about. Additionally, we need to specify our concept of cognition, in order to clarify the definition of learning and its attributes. Following, we offer a variety of empirical examples to illustrate what we have considered as plant learning.

2 Epistemological Framework and the Systemic Concept of Plants

Indeed, we are not totally convinced that the term "neurobiology" is the best option to name the new scientific perspective of the plant realm. It is not because we think that metaphors (if, indeed, is the case of the prefix "neuro") and analogies are not sufficient to support a science, rather because we think the term is actually restrictive to account for the whole complexity of the phenomena that it have claimed. Overall, the object of study of the science of "plant neurobiology" is neither the plant itself nor some specific structure or process of the plants; such is in classical human neuroscience. Instead, the object of the study is the relationship plant-environment. Thus, the main problem is to understand how plants survive and develop embedded in a changing world, creating stable communities (physiognomies) and evolving. Regarding the epistemological framework, which will be discussed below, the study of such relational object demands an interdisciplinary, even transdisciplinary, science (Barlow 2008; Debono 2013a, b). Accordingly, in the recent manifesto by Paco Calvo (2016), it proposed the "hexagon of plant neurobiology", showing the relationships among different domains of the human knowledge, representing the interdisciplinary collaborations (philosophy, ecology, (electro)physiology, cellular and molecular biology, biochemistry, evolutionary and developmental biology), supporting plant neurobiology studies. Coincidently, the set of collaborative sciences proposed by Calvo (2016) is essentially the same that constitute the classical science of physiological plant ecology (often named ecophysiology), which began early in the eighteenth century, with the same object of study: the plant-environment relationship in different scales of organization (Larcher 1995; Lüttge 2008).

However, throughout the development of ecophysiology, concepts related to aspects of plant cognition have not emerged in the theories and hypothesis to explain the plant-environment relationship. Why is it so? In my opinion, as a plant ecophysiologist, despite the intrinsic more holistic perspective, ecophysiology was born and has grown in the world dominated by the modern scientific paradigm based on both the mechanical principles and the reductionist view. Therefore, although concerning a very complex object of study, the underlying understanding (even unconsciously) that plants are reactive organisms (rote behaviour), instead of cognitive systems, has been dominant. Additionally, there is a strong and still growing influence of the molecular biology approach, supporting the belief that all biological phenomena can be explained by the processes of low level in the cells, the "omics" era (Sheth and Thaker 2014). Nevertheless, concomitant with the development of plant neurobiology, there is a solid criticism emerging inside the traditional plant ecophysiology proposing an explicit systemic perspective of this science, even though not attributing cognitive abilities for plants (Lüttge 2012; Matyssek and Lüttge 2013; Souza et al. 2016a).

Notwithstanding, recently Souza et al. (2016b) have assumed a link between the "systemic" plant ecophysiology and the perspective of plant cognition claimed by plant neurobiology. Thus herein, in order to distinguish from the current "systems biology" view, we have assumed a "systemic" perspective. We think this differentiation is necessary because "systems biology", despite a more holistic view of life processes, is restricted to the cellular domain (Sheth and Thaker 2014) and, thus, still reductionist, while the "systemic" approach is derived from the general systems theory (GST) (von Bertalanffy 1968) based on principles of multiscaled self-organized systems (Souza et al. 2016b). This approach is not essentially different from the perspective proposed by Barlow (2008) based on the living systems theory (LST) (Miller 1978). Actually, the main difference is that LST assumes an explicit cybernetics/biosemiotics approach describing any living being, wherever the scale of organization (form single cells to ecosystem), as input-output informational system composed by receptors, decoders and transducers. We are not convinced that such "analogical" simplification (actually a reduction to) to pure informational systems is adequate and sufficient to account for the complexity of all living beings, specially the cognitive aspects. One main problem in adopting an informational model of plant cognition is that such approach often falls in a type of "mental" representational cognitive system (Garzón 2007). As will be discussed later, this is not the unique model for the understanding of plants as cognitive systems. Thus, we have considered more appropriate to maintain the original epistemological position set on GST and its derived complex system theory (Mitchell 2009), which offers more possibilities to develop new insights and hypothesis. Moreover, in certain way, LST is included in GST.

Accordingly, we consider plants as non-equilibrium systems, open thermodynamically, in the sense of Nicolis and Prigogine (1989). Therefore, the organization of such systems relies on continuous flows of exchanges of matter, energy and information with the surrounding environment (Schneider and Kay 1994). Thus, plants must deal with the constant fluctuation of the environmental resources (light energy, water, nutrients) that eventually may constrain or threaten their stability (Souza and Lüttge 2015). The ability to face the complexity of the environmental changes is particularly critical for sessile organisms such as plants (Trewavas 2003, 2014).

Because of the limited capacity of locomotion, plants must deal with all sorts of environmental variation in their surroundings. As a result, their stability requires dynamic elements that confer some degree of organizational flexibility. To the set of possible changes in response to external stimuli, we call phenotypic plasticity (DeWitt and Scheiner 2004). Therefore, there is a close correlation between the stability of a system and the plasticity of phenotypic responses for a specific genotype. In essence phenotypic plasticity gives the system the ability to expand its capacities of physiological adjustments (in the case of individuals) or genetic adaptation (in the case of populations). The plasticity, particularly the modulation or control of plasticity, is related to patterns of organization of biological systems, which are characterized by complex networks (Hütt and Lüttge 2005; Souza et al. 2009). Overall, biological systems are essentially complex adaptive systems.

Notwithstanding the environment triggers responses in the organism, the pattern of such responses is determined by the internal dynamics of the system itself. This internal dynamics is integrated in a complex metabolic network that operates out of some rules of interactions. The rules that specify the interactions among the system components are performed using only local information, without reference to a pre-existing global pattern (Camazine et al. 2001). In complex systems, such interactions are typically nonlinear processes based on negative and positive feedback loops. The negative feedback plays a crucial role in maintaining homeostasis of the system, whereas the positive feedback operates propagating and amplifying signals throughout the system. Both processes work together in the formation and stabilization of new patterns of organization, which makes difficult the prediction of their global behaviour (Camazine et al. 2001; Lüttge 2012). Such a dynamical process of organization, operating through the different scales of the organization of the system, produces emergent properties and, hence, makes the system as whole non-reducible to its components at smaller scales of organization (Souza et al. 2016a)

Besides the complexity inherent to the internal dynamics of the system organization, there is the effect of environmental noise (random fluctuations of physical factors of the environment) that interferes in plant responses to specific environmental factors (Bertolli and Souza 2013). Indeed, a simple combination between two or more stress factors is capable to change significantly the plant responses (Prasch and Sonnewald 2015). Henceforth, it is clear the huge challenge for plants to survive in such complex world, where simple reactive strategies to solve all sorts of constraining faced by plants in their environment, would be unlikely. According to Trewavas (2003, 2005), only organisms with intelligence (the capacity for solving problems in an efficient way) would be able to support its organization and, in consequence, maintain the fitness.

Another remarkable characteristic in plants, defining part of their own existence, is the modular structure of their bodies. A module can be considered as a biological entity (an individual, a structure, a process or a pathway) characterized by more internal than external integration (Bolker 2000). Overall, modules can be considered as the knots of networks that are connected via the edges (different modes of short- and long-distance signalling). In hierarchical networks (such as any biological system), networks of finer scales can become knots in networks of higher scales and so on (Lüttge 2012). In plant biology, de Kroon et al. (2005) use the term module to refer to "repeated, often semiautonomous, structural and functional units". According to de Kroon and collaborators, "the response of a plant to its environment is the sum of all modular responses to their local conditions plus all interaction effects that are due to integration". In other words, taking the definition of plants proposed herein, the responses of plants to environment are emergent properties. Emergence is the inevitable self-organized unfolding of new functions and structures of a system on a higher scalar integrative level (Lüttge 2012). This is a far-reaching assumption. It is implied from it that, for instance, physiological responses (from the point of view of the external observer) are emergent properties from the lower organization level (cellular/molecular level); therefore, such responses cannot be accurately inferred from below. In a more clear word, ecophysiological responses to environmental cues are not reducible to molecular phenomena in a straightforward way, although they have correlation (taking into account that correlation does not imply necessarily causation) (Vítolo et al. 2012; Bertolli et al. 2014; Souza and Lüttge 2015; Souza et al. 2016a).

Modularity and emergent properties bring, at least, two challenges to the understanding of the plant learning abilities: (1) Who does it is the "individual" that learns? and (2) How and where to observe the learning? Howsoever, both challenges fall in the problem of individuality in plants.

Gilroy and Trewavas (2001) have hold that individuality is a term used to describe sets of structures morphologically similar (cells, tissues or plants) showing unique responses to signals. Overall, the expression of heterogeneous behaviour in a set of similar structures indicates individuality. For instance, the heterogeneity observed in the stomatal responses in a single leaf indicates that each single stoma responds in a different way under a specific situation, and eventually, a group of single cells can synchronize exhibiting a local collective behaviour (Mott and Buckley 2000). Analogously, the phenology of trees subjected to different environmental changes may exhibit complex spatiotemporal patterns with likely far-reaching consequences for ecosystem and biosphere functioning and structure (Peñuelas et al. 2004). Therefore, according to Clarke (2012), the individuality in plants is a matter of degree, depending on the hierarchical scale of observation. Some plants, in some circumstances, give us reason to say that modules are individuals, while other plants will exhibit different properties and would be best viewed as having individuality at a higher level.

Thus, somehow, individuality in plants, as modular organisms, depends on the scale of observation chosen by the observer himself, recognizing patterns of responses in groups of similar structures in a certain hierarchical level of organization. Howsoever, plants could be seen as networked multi-agent systems (Olfati-Saber et al. 2007) insofar the semiautonomous modules, connected by a complex network of different types of short- and long-range signals (Baluška et al. 2006; Baluška 2013), may engender a dynamic of consensus and collaboration, resulting in an expression of higher-level individuality, at whole plant level.

A possible explanation for individuality is the occurrence of stochastic processes during development (Gilroy and Trewavas 2001). The biochemical process underlying cell development (cellular, division, differentiation and growth), in particular the role of the enzymes in biosynthesis, results in a complex nonlinear dynamics, since a single enzyme may function in many interconnected enzymatic pathways. Moreover, the diffusion and the number of signalling molecules (e.g. plant hormones) among developing cells follow a complex self-organized non-homogenous dynamics, inducing minimal differences among cells. Nonlinear complex systems, such as cells' whole metabolism, are sensitive to small perturbation, mainly under suboptimal conditions. Thereupon, random perturbations tend to become magnified, increasing the differentiation among groups of cells and creating independence from each other (Møller and Swaddle 1997). Moreover, the complex electrical network, named "electrome", underpinning cellular activities, shall display a major role on the organism individuality. De Loof (2016) holds that there are not two identical electromes. Each cell has its own electrome that is different from any other cells, even considering two cells derived from mitosis. This is reasonable insofar the daughter cells have different distributions of quantities and types of ionic channels and pumps, as well as different cytoskeleton structures, which coordinate the fluxes of ions generating electrical activity (Debono 2013a, b; De Loof 2016).

Such "epigenetic noise" allows a variation of responses, from the cell to the population level to a plethora of environmental cues; thus, individuality forms a basis for phenotypic plasticity (Gilroy and Trewavas 2001). Plasticity is one of the bases of plant stability, conferring multifunctional regulatory capacity for plants (Souza and Lüttge 2015).

Finally, in order to define plant learning both in theoretical and practical terms, it is necessary to make explicit the concept of cognition that underlies our understanding.

Summarizing:

1. Learning in plants (as well as other cognitive abilities) shall be considered under the light of a more holistic scientific paradigm, and we thought that the systemic approach accomplishes this task. Discussing such likely abilities with the lens of the classical paradigm (mechanist and reductionist) is unfruitful and makes nonsense insofar the epistemological bases and the overall understanding on the object of study are radically different.
2. Plants are open systems, and their entire organization depends upon such openness. Thus, the object of study is the plant-environment system.

3. The modularity of plant's body organization makes difficult to define a single scale of observation representing the "whole plant". Therefore, the definition of the individuality itself depends on the observer and upon the contingencies around it.
4. Individuality is the basis for plasticity and, in turn, forms the basis for plant surviving in a changing environment.

3 A Framework for Plant Cognition

Insofar as learning is one aspect of the cognitive phenomena (the main target in plant neurobiology) (Calvo 2016) and there are different approaches in the cognitive sciences (Gomila and Calvo 2008), it is worthy to present a general framework, on which we stand a position in order to define "plant learning".

We think two main aspects are critical to be accounted: (1) the concept of representation (or no representation at all) and (2) the extension that cognition can reach in the natural world (beyond a human capacity).

Classical cognitivism is founded on the perspective of mental representations (modern philosophy) and information processing systems (artificial intelligence, computational neuroscience). In this traditional approach, the cognitive capabilities of the mind are conceived as disembodied, and the cognitive phenomena are trivially depended upon the environment in a stimulus-response basis. Influenced by the cybernetic approach, the metaphor of the brain as hardware and mind as the software exacerbates the Cartesian dualism on which it is based on. Symbol structures are assumed to correspond to real physical structures in the brain, and the combinatorial structure of a representation is supposed to have a counterpart in structural relations among physical properties of the brain (Gomila and Calvo 2008; Richardson et al. 2008). In consequence, usually, cognition is thought as it were predominantly a human faculty and applied to just one level of organization (the organismal) (Barlow 2010; Cazalis et al. 2017).

On the other hand, in the *post-cognitivism*, cognition is considered not as an abstract computation; instead cognition is viewed as interactive, embodied and embedded system. Despite the many research programmes in post-cognitivism, the different approaches take cognition and behaviour in terms of the dynamical interaction of an embodied system (a real biological organism, for instance) that is linked to the surrounding environment (embodied-embedded nature) (Gomila and Calvo 2008). In this way, cognition incorporates the notion of an environment with an active role, instead of a passive one in the organization of the organisms (Cazalis et al. 2017). This is a major step towards a concept of cognition extensible to plants insofar as plants have a modular (non-centralized) body inextricably connected to its environment. Under this perspective, plant is taken as an *individual-coupled-with-its-environment system*, and cognition is rather an emergent and extended self-organizing phenomenon. Centralized processes are not necessary (Garzón 2007).

Accordingly, the "Santiago theory of cognition" developed by Maturana and Varela (1980) holds that living systems are cognitive systems, and living is a process of cognition, which applies irrespective of the presence, and intervention of a nervous system. Accordingly, any organism by definition is the manifestation of cognition, since this is the way by which it can be a social entity that can experience the environment (Cazalis et al. 2017). The concept of cognition as a basic biological phenomenon is assumed, as a general assumption, by plant neurobiologists such as Brenner et al. (2006), Garzón (2007), Barlow (2008, 2010) and Gagliano (2015). Garzón and Keijzer (2011) summarize this general standing: "cognition is a biological phenomenon, and that it exhibits itself as a capability to manipulate the environment in ways that systematically benefit a living organism".

Thus, at least under the perspective of post-cognitivist, it is reasonable to assume cognition as a property/process extensible to plants. However, the question about what position, representational or non-representational, would be more appropriate for a model of plant cognition remains.

Broadly, in the perspective representationalist, cognitive systems are taken as information processes that produce representations, which can exploit it in a purposeful manner; on the other hand, the anti-representationalist framework holds a non-computational way to understand the relationship between an organism and their environment (Calvo 2016). In order to take a cognitive system as representational, when the processing of representational states marks cognitive activity, Garzón (2007) has proposed two main principles: the principle of dissociation, meaning that a physical representation must stand accessible even when things or events are not available temporally, and the principle of reification, whenever the representational states are clearly identified to a computational role they are supposed to play.

According to Calvo (2016), a problem with applying the representationalist approach to plants is that the perception is the outcome of a logic-like process of inference, because the stimulus is "inherently ambiguous". Recently, an alternative non-representationalist and antimechanist view has been revisited: Gibson's ecological psychology. Although originally Gibson's framework (1966) was restricted to animal behaviour, his ideas have been extended to plants (Gagliano 2015; Calvo 2016). Gibson argues that "animals" should not be conceived as machines, and the mechanistic view of the environment as a matter in motion is inappropriate to understand animal behaviour. Instead, Gibson holds that the environment of the animals consists of action possibilities, which he termed *affordances* (Withagen et al. 2012).

Two aspects of the concept of affordance are remarkable: (1) environment itself is meaningful, and (2) environment consists of opportunities for action, i.e. environment is not conceived as a collection of causes but as a manifold of action possibilities (Withagen et al. 2012). Under the ecological framework, perception is organized around action. Opportunities for action could be "perceived directly" as interaction with an "unambiguous environment". Thus, plants perceive opportunities for behavioural interaction in the form of affordances (Richardson et al. 2008). Affordances are not properties of the phenomenological world that depend upon the state of the observer; rather, they are ecological phenomena that exist in the environment. Thus, environment does not cause behaviour but simply makes it possible (Withagen et al. 2012).

At least, two major problems can be considered in the ecological psychology approach: the supposed unambiguity of the environment and the direct perception. Actually, as we argue before, the natural environment of a plant is very complex with a plethora of stimuli affecting each other in time and space, and thus, it is inherently ambiguous. Accordingly, the interaction of the plant with the surrounding has a complex nonlinear dynamics, demanding an integration of many ambiguous signals in order to allow the stability of plant organization. This is specially a critical ability for a modular organism (Souza and Lüttge 2015). For instance, stomatal responses to simultaneous applied opposing environmental factors cannot be predicted from studying one factor at a time. The stimulation induces ambiguous responses that are species-specific (Merilo et al. 2014).

However, the concept of direct perception is more controversial, because its explanation is flaw. According to Richardson et al. (2008), "solutions, perceptually speaking, emerge out of the interaction between the organism and its local environment". There is no clue how such direct perception takes place. It is a hypothesis hard to be tested empirically.

Moreover, the enormous amounts of empirical data, showing well-known strong correlation between plant physiological changes and abiotic and biotic cues (Atkinson and Urwin 2012; Jenks and Hasegawa 2014; Azooz and Ahmad 2016), indicate that plant metabolism and, overall, plant behaviour are inextricably coupled with the environment (Trewavas 2009). Indeed, plant responses to changes in environmental conditions involve corresponding changes in the perceptual apparatus of plant cells. Such perceptual apparatus consists in specific membrane receptors for different signals that trigger cascades of signalling transducers that, in turn, regulate the gene expression and the corresponding metabolic changes. Enormous numbers of molecular connections integrate into a complex self-organized and dynamic network, modulating plant behaviour under changing external conditions (Sweetlove and Fernie 2005; Trewavas 2005; Lucas et al. 2011).

What we are suggesting is that such metabolic organization, coupled to environmental cues, is, indeed, a "molecular representation" of a corresponding environmental condition. Those "representations" are, indeed, changes in metabolic network topologies (different arrangements among cellular components) that engender metabolic schemes corresponding to the current status of the cells under certain local conditions. Such schemes can be stored and recalled later in certain circumstances (storage/recall type of memory), eventually changing future plant behaviour (Thellier and Lüttge 2012).

There is no plant physiological response that is not mediated by changes in the perceptual apparatus of the plants. Thus, "direct perception" is, in our opinion, a misleading and unnecessary concept without correspondence with empirical data from ecophysiological studies.

Notwithstanding the criticism on some aspects of ecological psychology, we agree with the perspective that environment does not cause behaviour but makes it possible. The changes expressed by metabolism are not ruled by environment, rather, they are emergent responses from the interactions among the networks of different modules in the plant body, integrating and canalizing external information

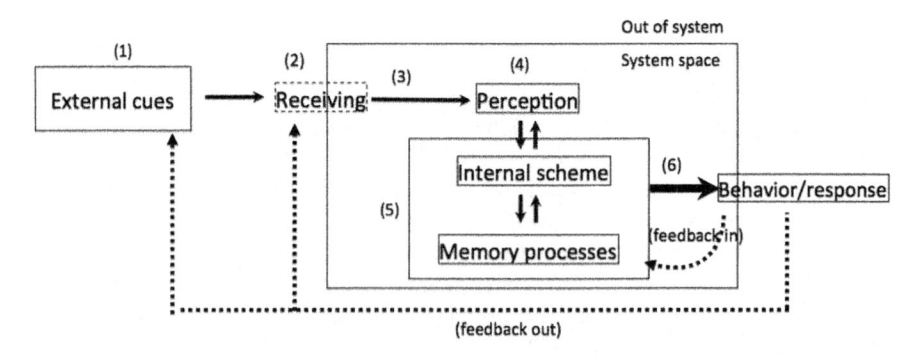

Fig. 1 General cognitive model. (1) Environmental factors (both biotic and abiotic); (2) plant membrane receptors (often proteins that recognize specific signals); (3) transducing cascade (often consisting in secondary signals, enzymes and transcriptional factors); (4) network metabolic shaping (including gene expression and further metabolic changes) engendering a cellular scheme; (5) feedbacks between storage and recall metabolic processes (memory); (6) integration processes producing local and/or global new behaviours (plant responses). The responses can modify directly the internal processes (feedback in) and/or affect external environment (feedback out)

into new states of organization, allowing plant stability. This perspective is consistent with the view of plants as autonomous individuals, instead of objects in a mechanistically conceived world.

Summarizing, the concept of "plant learning" developed below is conceived on the following cognitive perspective: plants are modular embodied-embedded systems, hierarchically organized, perceiving and acting on the world by manipulating local representations (cellular schemes) and by means of electric ionic waves guided by proteins and feeding back on the same proteins, determining the plasticity of the active signal transduction pathways and integrating them globally through short- and long-distance signalling processes.

As a consequence of organizational changes in the plant (structural and/or functional), its surrounding environment can be altered, refeeding the perceptual system of the plant and, eventually, generating new internal schemes and changes in the respective signalling pathways (Fig. 1). However, this process of true feedback between the plant and its environment can occur locally in specific modules. For example, during the development of a tree canopy, different leaves develop in different positions receiving different amounts of light energy with different durations and time interval (sunflecks), creating micro-environments (Watling et al. 1997). On the other hand, the perceptual process can occur in the whole individual, as in controlled experiments with tiny plants of *Arabidopsis thaliana* (Kreps et al. 2002). Whatever the case, a sophisticate system of internal signalling takes place in order to integrate local or global cellular changes (Choi et al. 2016).

Different types of signals, such as hormones, ROS, Ca^{2+} and electrical signals, compose the plant's signalling network (Baluška 2013; Choi et al. 2016). Increasing strong evidences have demonstrated that electrical signals play a central role in both cell-cell and long-distance communication in plants (Baluška et al. 2006; Fromm and Lautner 2007; Gallé et al. 2015). Electrical signal transmission takes place throughout cellular connections in the symplasmatic phloematic continuum, creating a complex network, similar to a simple neural net (Baluška et al. 2006; Debono 2013a, b; Choi et al. 2016), engendering the plant electrome (De Loof 2016). Evidence of synchronization of electrical spikes among different cells, indicating a collective behaviour of groups of cells interconnected by the plasmodesmatal network, supports the neuroid conduction hypothesis in plants (Masi et al. 2009; Debono 2013a, b). Recently, Saraiva et al. (2017) have demonstrated that electrome in plants, measured as time series of low-voltage variations, can exhibit high complex dynamic patterns (actually, chaotic behaviour instead of purely white noise). Moreover, the complexity of the electrical signals showed dependence of the environmental stimulation, exhibiting burst of electrical spikes following a power law (i.e. spikes without a characteristic size), suggesting that plant electrome can be critically self-organized (Souza et al. 2017). Such complexity inherent to the signalling network in plants can support a massive informational processing system, enabling plants to process information in order to keep their stability and learning processes.

3.1 Concept of Learning

As with cognition, there are different possible approaches to an adequate concept of learning in plants. Recently, an interesting and provocative contribution from comparative psychology and behavioural biology provides a scope to discuss possible aspects of learning in plants, based on studies focused in the species *Mimosa pudica*, the sensitive plant (Abramson and Chicas-Mosier 2016). They have proposed an empirical basis in order to test for associative and non-associative learning. Accordingly, non-associative learning has two categories: habituation and sensitization. Habituation refers to a decrease in responding to a stimulus that is relatedly presented, while sensitization can be considered the opposite of habituation since it refers to an increase in the frequency or probability of a response. In a recent study with *M. pudica*, Gagliano et al. (2014) showed that the leaf-folding behaviour in response to repeated disturbance exhibited clear persistent habituation, mainly when plants were grown under energetically costly environments.

On the other hand, associative learning has in conditioning its most basic expression, accessed by stimulus-response training experiments (Abramson and Chicas-Mosier 2016). Gagliano et al. (2016) have tested this hypothesis for plants in a clever set of experiments with *Pisum sativum* (garden pea), emulating the Pavlovian empirical approach. By using a Y-maze task, it was showed that, after a period of training, the position of a neutral cue (a fan located in alternated positions in the arms of the Y-maze), predicting the location of a light source, affected the direction of plant growth, prevailing over innate phototropism.

Despite some promising empirical data, we are not sure about the ubiquity of such perspective throughout the plant kingdom, as well as on its role for plant living. Actually, maybe the behavioural approach is not consistent with the plants' general lifestyle. According to Trewavas (2014), "unlike animals, plants are not unitary organisms; they constantly change throughout their development process: we do not deal with the same plant twice". The attempts to transpose the framework from the psychology behaviour to the context of plant learning (Abramson and Chicas-Mosier 2016) present some constraints that, in our opinion, are not suitable for plants' living style. First of all, plants do not exhibit movements of their whole bodies (unitary individuals) from one place to another (A-to-B movement), like the free movements of animals. Despite the different types of tropisms, allowing plants to explore their surroundings, these types of movements are *not* actual free movements. Plants are typically sessile organisms, and all sorts of evolutionary adaptations match with their lifestyle. Thus, under the point of view of a plant, A-to-B movements simply are not necessary. Even taking the growth of plants as analogous to animals' "free movements" in order to adapt experiments from psychology behaviour to test for the different categories of associative and non-associative learning is, somehow, distant from the plant lifestyle in the wild.

Second, as we have discussed lately, there is the problem of individuality in plants. Plants are not typical unitary individuals. Notwithstanding the many modules that communicate with each other by a complex network of signalling, allowing some level of integration, each module is semiautonomous and, under certain circumstances, behaves like an individual. Furthermore, under natural conditions, where plants face an enormous complexity of environmental cues in different scales of space and time, the individuality of the modules is exacerbated (de Kroon et al. 2009). Experiments carried out under very specific and controlled conditions, although they may give some insight into possible behaviours, are far removed from the complex reality of plants (Trewavas 2003, 2014). Accordingly, Garzón and Keijzer (2011) hold that minimal forms of cognition, be it plant, bacteria or animal, cannot be studied detached from the natural habitat in which they take place; insofar under laboratory conditions, the selection pressures are manipulated, affecting the observed behaviour. For instance, the capacity of soybean plants to respond to water deficit was different when plants were grown under environmental controlled conditions (inside a grown chamber) compared with the plants subjected to same water-deficit imposition, but grown under more variable conditions (Bertolli and Souza 2013). Recently, Vialet-Chabrand et al. (2017) have showed the importance of fluctuation in light on plant photosynthetic acclimation, showing that growing plants under square wave growth conditions ultimately fails to predict plant performance under realistic light regimes. Therefore, we believe that a more broad and general concept of learning could afford a more suitable framework for the plant learning understanding.

First of all, learning is not an all-or-nothing property; there are levels of learning. The concept of learning levels (L0 to L3) was conceived by Gregory Bateson (Bateson 1972), insofar as organisms can become responsive to patterns, but then can become recursively responsive to patterns within those patterns. Herein, according to Affifi (2013), we have hold that plants show, at least, the L1 level, which living beings can achieve without a neural circuit: these are the cases in which an entity gives at Time 2 a different response from what it gave at Time 1 to a "similar stimulus". Cleverly, Bateson holds that the assumption of the repeatable stimulus is critical for the concept of learning; insofar in the wild, the existence of exactly the same circumstances in Time 1 and Time 2 is very unlikely. Otherwise, no learning would take place, and all responses fall in the level L0 (when an entity shows minimal change in its response to a repeated item of sensory input).

Whatever, according to Bateson (1972), learning undoubtedly denotes change of some kind, and change denotes process. Therefore, learning is process. Such perspective matches with the concept of learning in Maturana and Varela (1980), where learning is taken as a process that consists in the transformation through experience of the behaviour of an organism. This transformation is a historical process such that each mode of behaviour constitutes the basis over which a new behaviour develops, either through changes in the possible states that may arise in it as a result of interaction or through changes in the transition rules from state to state. Learning occurs in a manner that, for the observer, the learned behaviour of the organism appears justified from the past, through the incorporation of a representation of the environment (memory) that acts, modifying its present behaviour by recall; notwithstanding this, the system itself functions in the present, and for it learning occurs as a temporal process of transformation.

Such dependence on the observer claimed by Maturana and Varela is in confluence with the systemic perspective of the nature insofar the observer defines the spatial and temporal scale of the observed phenomena and, consequently, defines the system itself under observation. A semiotic interpretation of learning (Affifi 2013), where learning is concerned with changes in sign activity, holds that besides levels of plant learning, in a *sensu* Bateson, there are also scales of learning, referring to the different spatial and temporal ranges within which learning can be studied (dependence on the observer). Accordingly, there are four major scales in which it is conceivable that plants learn. On the first scale, learning occurs by parts of the plant (modules or individual cells) being regulated by the activity of the plant as a whole. The other levels involve the whole plant as an individual (second scale) and learning at the level of population or community (third scale), by processes of plant-plant communication; and the fourth scale includes the species scale, and through natural selection, a species "learns" certain adaptive behaviours. Affifi (2013) has admitted that in the fourth scale, "learning" takes on a very different sense, no longer associated with anything potentially experiential or phenomenal.

No matter what the scale is, learning is an emergent property from the interactions among elements of the system (or part of it) involved in the cognitive process. According to Trewavas (2003, 2014), learning process involves construction of networks, such as plasmodesmatal connections among cells, allowing exchange of information among cells through fluxes of proteins, nucleic acids and other smaller molecules, as well as new tissue formation, giving plants the flexibility to changes in their phenotype in a changing environment (phenotypic plasticity). Furthermore, plasticity is also observed experimentally by means of electrical registers of plant tissue activity (Debono 2013a, b; Souza et al. 2017) generated by ionic fluxes in solution, which are guided by proteins (membrane channels and pumps). Thus, considering that plant tissue is a symplasm that allows the intercellular flow of ions and small molecules, engendering a complex electrical network, the learning mechanisms should be looked for in this context, in a similar way as neuroscience.

Spontaneous activity of membrane variation potential is an important property of cerebral cortex that is probably involved in information process and storage (Arcangelis and Herrmann 2010). Neuronal avalanches in spontaneous spatiotemporal activity were already shown in very few to very large numbers of neuronal assemblages in organotropic cultures of rat cortex slices. The size and duration of avalanches in such neuronal bursts of firing follows a power law distribution with very stable exponents (Beggs and Plenz 2003, 2004). The power law distribution is a signature feature of systems activity under critical state, interpreted in all cases as self-organized criticality (SOC) (Bak 1996; Arcangelis and Herrmann 2010). Timely, recent results of temporal dynamic of low-voltage electrical signals in plant tissues have showed evidence of SOC behaviour. Under different environmental stimuli, Souza et al. (2017) have observed appearance of electrical spikes following a power law. Emergence of SOC is associated with mechanisms of slow energy accumulation and fast energy redistribution, which drives the system to critical state, where avalanche extensions and durations do not have a characteristic size (Bak 1996; Arcangelis and Herrmann 2010). The ability of the brain to self-organize connections has been part of theoretical models for learning and verified in real brain activity, where operating in critical level, far from uncorrelated subcritical or too correlated supercritical regime, optimizes information management and transmission (Arcangelis and Herrmann 2010; Beggs and Plenz 2003; Kinouchi and Copelli 2006; Eurich et al. 2002). In neuronal models of learning, acting in critical state, where a reproduction of physiological mechanisms of neural behaviour is implemented on a network with topological properties similar to brain network, shows how learning and memory can occur in function of different levels of plastic adaptation introduced via non-uniform negative feedback (Arcangelis and Herrmann 2010). The same model also demonstrates that learning dynamics shows universal properties, independent of the details of the system or the specific task assigned, since slow plastic adaptive parameters are used. The learning process seems to be very sensitive to initial conditions using the model (Arcangelis and Herrmann 2010), which may be indicative of chaotic traits (Schroeder 1991). Similarly, Saraiva et al. (2017) have showed some empirical evidence for the whole plant electrical signals approaches a chaotic dynamic as well.

The conclusions of testing the neuronal model are that systems act in critical state response to given inputs in high flexible way, adapting itself more easily to different learning rules, and that learning is a collective process dependent on connectivity level and on plastic adaptive strength parameters. The requirement of slow plastic adaptation was confirmed in experimental analysis in humans, showing that learning performance improves when minimal changes occur in functionality network (Shew et al. 2009). So, the learning process is dependent on plasticity in neuron function and connectivity, which is an example of phenotypic plasticity. Accordingly, based on the evidences of SOC behaviour in plant electrophysiology (Saraiva et al. 2017;

Souza et al. 2017), it would be worthy to perform more studies testing the hypothesis that plant learning may be supported by a self-organized electrical network approaching the critical state.

No matter what scale of organization is under consideration, it is clear that some level of phenotypic plasticity is the basis for plant learning. According to Trewavas (2003, 2014), learning involves goals and error-assessment mechanisms that quantify how close the new behaviour approaches that goal (mostly the fitness). Thus, the process of learning requires a continual exchange of information and feedback from the goal to the current behaviour in order to correct current behaviour and direct future more closely towards achieving the goal, such as illustrated in feedbacks represented in the model showed in Fig. 1. It is equivalent to a trial-and-error process that can be indicated by the presence of robust oscillations in behaviour as the organism continually assesses and makes further correction to the behaviour. Stomatal complex oscillations in response to changes in environmental water conditions are nice examples of timely dynamical behaviour (Souza et al. 2004, 2005). Accordingly, phenotypic plasticity allows the error-correcting mechanisms that individual plants use in an attempt to achieve optimal fitness (Trewavas 2003).

The ability of plant to learn can be grasped from acclimation and/or optimization behaviours that allow plant to survive and, eventually, maintain or even increase its fitness under certain circumstances. In a working definition proposed by Karban (2015), "learning requires that past events cause chemical changes that influence the sensitivity, speed, or effectiveness of subsequent plant sensing and associated responses".

In order to hold a clear cut between acclimation and learning, we propose that learning shall consist in phenotypic changes based on memory that, effectively, improves plant responses to similar stimuli experienced in a certain moment in the past and then it was ceased (no longer accessible for the plants). In situations that environmental changes represent a new and permanent condition, the responses of plants are also continuous allowing plant acclimation (according to the *Merriam-Webster Dictionary*, "*acclimation* is the process or result of acclimating; specially: physiological adjustment by an organism to environmental change"). This concept of learning is aligned with our representationalist approach, matching the principle of dissociation proposed by Garzón (2007) that, plus the principle of reification, supports the processing of representational states. Following, we offer a variety of examples to illustrate likely situations that learning could be claimed.

4 Empirical Evidences

4.1 Experiencing Biotic Stimuli

Co-evolution with a range of pathogens and insects has enabled plants to evolve to defend themselves against great part of injurious microbes, insects and other parasites. Plants commonly react to attacks by producing defensive molecules and compounds that are toxic to the aggressor. As a co-evolutionary response, plants acquired the capacity to recognize hostile and virulent pathogens, even at early stages of infection or attack, inducing an appropriate defence response. Such ability involves the capacity to store information (memory) of the experiences of previous interactions with other organism and, then, uses them for improving further responses, a phenomenon known as "priming" by researchers studying abiotic stress (Crisp et al. 2016).

The phenomenon of priming is a common topic underlying responses to a range of biotic stresses, wherever previous exposition to some pathogenic organism makes a plant more resistant to future exposure to a stressful event. When a plant is "primed", the information of the priming stimulus is stored until exposure to a triggering stimulus in the future, an effect known as "memory" in plant defence (Kandel et al. 2014). Primed plants many times show longer-lasting activation or attenuated repression of defence upon challenge than unprimed plants (Hilker et al. 2016). Defensive traits that can be primed may include, among others, changes in defence-related signalling compounds or processes such as hormones and enzymes, alterations to chromatin or enhanced presence of pattern-recognition receptors or defence responses, such as accumulation of phytoalexins, phenolic compounds, reactive oxygen species, glucosinolates, lignin or plant volatiles induced by herbivory (Balmer et al. 2015). Thus, primed plants show either faster or stronger activation of the several defence systems. Therefore, the responses that are induced following attack by either pathogens or insects facilitate a more rapid response if the stress recurs (Conrath et al. 2006; Ton et al. 2007). It is important because when plants anticipate herbivory, through the perception of indicative signals or by experience of herbivory at their parental generation, plants are physiologically prepared to induce stronger and faster defences upon the anticipated stimuli. Moreover, this strategy may also reduce the possibility of development of counter-adaptive strategies by insect herbivores to increased plasticity of induced defence by priming (Zheng and Dicke 2008). Therefore, plants have some form of "memory" often termed "stress imprint", such as a genetic or biochemical modification of a plant that occurs after stress exposure that causes future responses to future stresses to be different.

There are studies on priming in the plant-insect interactions since the first report of defence in maize (Engelberth et al. 2004). Currently, priming research consists in studying priming defence mediated by herbivore-inducible plant volatiles (HIPVs), which are produced and released by the neighbouring plants, or plant parts, under herbivore attack. In literature, a number of experiments have shown that priming effects can last for several days at least. For example, tobacco plants can store information on previous induction for at least 6 days, and in *Arabidopsis* plants, such information can be stored at least 3 days (Goh et al. 2003).

HIPVs' mechanisms are important to plant defence, because it occurs many times in the plant life. HIPVs are green leaf volatiles [GLVs], terpenoids and other molecules, actuating as plant signalling molecules (between and within plants) or attracting natural predators of the herbivores (McCormick et al. 2012). HIPVs can induce or prime defensive responses in neighbouring intact plants or intact plant parts on the same plant. Several volatiles and GLVs produced in response to insect feeding and mechanical damage are capable of priming defences (Heil and Kost 2006; Choh and Takabayashi 2006; Kessler et al. 2006; Heil and Silva Bueno 2007; Frost et al. 2007; Ton et al. 2007; Rodriguez-Saona et al. 2009; Muroi et al. 2011; Peng et al. 2011; Hirao et al. 2012; Li et al. 2012). The primed defences consist of different processes: accumulation of jasmonic acid (JA), inducing anti-herbivore defences (Engelberth et al. 2004; Frost et al. 2008) and accumulation of linolenic acid, a precursor of JA and GLVs (Frost et al. 2008); increased production of plant secondary metabolites (Kessler et al. 2006; Hirao et al. 2012); increased protease inhibitor activity (Kessler et al. 2006); and enhanced transcription of anti-herbivore defence genes (Ton et al. 2007; Peng et al. 2011). Furthermore, increasing emission of HIPVs and secretion of extra floral nectar, an extra sugar source, attracts predators of plant enemies, such as ants, improving natural plant defences (Heil and Silva Bueno 2007; Rodriguez-Saona et al. 2009; Muroi et al. 2011; Li et al. 2012; Ton et al. 2007; Peng et al. 2011).

Mechanisms of defence priming in plants are caused by signals that indicate attack by pathogens or herbivores. This includes the known "systemic acquired resistance" (SAR), which is triggered by pathogen attack and causes a systemic priming of salicylic acid (SA) that induces defence mechanisms (Jung et al. 2009; Kohler et al. 2002). Another example of stress-indicating priming signals is volatile organic compounds (VOCs), which are released by herbivore-infested plants. Several VOCs can prime JA-dependent defences in plant modules and/or neighbouring plants (Turlings and Ton 2006; Ton et al. 2007; Heil and Ton 2008).

Nevertheless, not all priming responses are triggered by adverse signal. For example, plant-beneficial organisms, such as non-pathogenic rhizobacteria and mycorrhizal fungi, can trigger priming that results in an "induced systemic resistance" (ISR) response (van Wees et al. 2008). Priming related to ISR is associated with priming of JA-dependent defences, because it is most effective against pathogens that are resistant to JA-inducible defences (Verhagen et al. 2004; Pozo et al. 2008; Ton et al. 2007).

There are many evidences showing that several stress-induced effects in plants can be transmitted to the next generation. Molinier et al. (2006) have observed genomic changes (hyper-recombination in the somatic tissue) of plants exposed to UV radiation and flagellin, an elicitor derived from bacteria, as well as in their non-treated progeny. The transgenerational stress "imprint" effects were also observed in wild radish (*Raphanus raphanistrum*) responses to herbivore damage by butterfly (*Pieris rapae*) and in plants treated with JA (Agrawal 2002). Accordingly, the progeny of treated plants were more resistant to herbivory than control plants. Another study showed that *Arabidopsis* plants exposed to localized infection by *Pseudomonas syringae* produce progeny more resistant to pathogen (Slaughter et al. 2012). This transgenerational effect is evident in progeny of plants that received repeated infections with *Pseudomonas*. Rasmann et al. (2012) showed that treatment with JA or exposition to insect herbivory makes *Arabidopsis* and tomato produce progeny more resistant against caterpillar feeding. Additionally, some studies reported an enhanced anti-herbivore resistance in plants whose parents experienced previous interaction with herbivores, for instance, in *Raphanus raphanistrum* (Holeski 2007), *Mimulus guttatus* (Holeski et al. 2012) and *Taraxacum officinale* (Verhoeven and van Gurp 2012).

Overall, the different examples above of plant biotic interactions illustrate quite well the ability of plants to grasp environmental information, engendering differing metabolic schemes corresponding with the different stimuli, and improve their responses in future interactions, an actual learning phenomena. Additionally, it is remarkable the possibility of transgenerational memory, improving behaviour of the next generation, an interesting possibility for learning at a population level (Affifi 2013).

4.2 Experiencing Abiotic Stimuli

In the literature of abiotic stress, the term "acclimation" is equivalent to "priming" in biotic stress studies (Crisp et al. 2016). However, different experimental designs are put together under the same idea of "acclimation". Many studies on the effects of abiotic factors on plant performance are carried out applying continuous stimulation over a certain period of time, and the plant states are compared before and after stimulation (e.g. Vítolo et al. 2012; Bertolli et al. 2014), while other studies are carried out with subsequent environmental stimuli (e.g. Kron et al. 2008). The types of acclimation that can be seen as learning are those examples whose stimulus is given; then it is cessed for a while, and when applied again, it leads to a more efficient response (more rapid and/or more intense). This kind of acclimation is also named as priming, such as for biotic studies. Thus, acclimation that happens continuously and gradually along time cannot be considered (or recognized with) learning if they do not have time interval between stimulus experience cessation and restimulation, because it is not possible to verify enhanced performance after an interval (more rapid and/or intense responses). Despite this, we also cannot say that continuous acclimating does not involve learning process at all.

Examples of abiotic priming are scarce compared with biotic priming, because priming concept was consolidated with abiotic stress studies (Conrath et al. 2006). An early example of abiotic stress priming was presented by Knight et al. (1998). *Arabidopsis* plants pre-exposed to osmotic and oxidative stress showed improved tolerance to subsequent stimuli compared with control. This behaviour was correlated with different patterns of Ca^{2+} signalling and genetic expression. Similarly, *Arabidopsis* plants pretreated with BABA (beta-amino-butyric acid) enhanced activation of plant defence system under post-drought stresses, and in this case "priming" was used to designate such response improvement after an interval of non-stimulating period (Jakab et al. 2005).

This phenomenon has been observed in other species as well. *Arrhenatherum elatius*, for example, displays different performances in response to different previous drought (Walter et al. 2011). *Brassica juncea* L. treated with hydrogen peroxide (H_2O_2) improved drought stress tolerance after 48 h (Hossain and Fujita 2013). There are many other examples using H_2O_2 priming to enhance stress response that can be found in Hossain et al. (2015). Another mustard species (*Brassica campestris* L.)

pretreated with 6° C cold shock improved seedling tolerance to salt and drought stress by modulation of antioxidative and glyoxalase system (Hossain et al. 2013a). Heat-shock pretreatment also improves antioxidative system of plants of same mustard species (*Brassica campestris* L.) when exposed to salt and drought stress (Hossain et al. 2013b). Nitric oxide (NO) is an ubiquitous and important signalling molecule that has shown a priming ability, triggering improved antioxidative responses to salt stress. Low doses of NO, or H_2O_2, applied in rice seedlings allowed higher survival of green tissues, and higher quantum yield for photosystem II, of plants post-submitted to both salt and heat stress. Further examples on abiotic priming can be found in an up-to-date overview by Filippou et al. (2012).

Studies on abiotic priming in seeds have shown long-lasting effects persisting during and after germination. This issue is well documented and discussed in reviews by Jisha et al. (2013) and Paparella et al. (2015). A remarkable example of seed priming was presented by Iqbal and Ashraf (2007). Seeds from two different cultivars of wheat were pretreated with different solutions of $CaCl_2$, KCl and NaCl then, when the seeds were subjected to a saline stress condition, showing a significant increase in tolerance compared with the seeds without pretreatment. Similar response was observed with tomato seeds pretreated with NaCl, exhibiting higher performance under a subsequent treatment with saline stress, mainly in advanced growth stages (Cayuela et al. 1996).

So far, herein, we clearly identify those examples as a learning process. Accordingly, other environment factors, as temperature, also can induce plant learning. Rice seeds primed with chilling presented less negative effects caused by post-chilling stress after germination, compared with non-treated seeds (Hussain et al. 2016a). Another experiment tested how different seed priming stimuli (hydropriming, osmopriming, redox priming, chemical priming and hormonal priming) could better enhance chilling tolerance in adult plants. Hormonal pre-stimulation with salicylic acid and selenium was efficient to improve plants' performance to post-chilling stress (Hussain et al. 2016b). Priming rice seeds of three cultivars with BABA also improved drought and salinity stress tolerance of seedling, mainly in the drought-tolerant cultivar in Vaisakh and in the salt-tolerant cultivar in Vyttila (Jisha and Puthur 2016).

There are several other abiotic examples of priming that are hidden behind "acclimation", "recovering" or "memory" terminology which could be related with learning (Trewavas 2005, 2009; Crisp et al. 2016). Anderson et al., for example, revealed that maize seedlings (a cold-intolerant species) treated with 14 °C (a moderate cold) 3 days before different chilling stimulus degrees showed an improvement in survival (79% for pretreated against 22% for non-treated) and development compared with the non-treated control plants. The maize pretreatment with hypoxia improved their survival time in subsequent anoxic incubation (Xia and Saglio 1992), and mustard seedlings pretreated with salicylic acid or high temperature (45 °C for 1 h) improved thermotolerance in subsequent heat shock of 55 °C (Dat et al. 1998).

Many studies on "cross-tolerance" or "crosstalk" have been reported (Capiati et al. 2006; Mittler 2006; Pastori and Foyer 2002; Artetxe et al. 2002; Mateo et al. 2004). These are interesting phenomena that show a kind of generalization in plant responses to environment factors, which is one of the learning properties (Abramson and Chicas-Mosier 2016). Generalization was a learning aspect tested in recent research that investigated learning in plants (Gagliano et al. 2016).

The cellular complex metabolic networks, more specifically the redundancy traits in some metabolic signalling pathways, enabling crosstalk and the consequent cross-tolerance, are evidence that complex metabolic networks have semantic properties (Witzany 2006), allowing plants to respond to stimulus in both generalized and specific ways. It is also important to emphasize again that in nature the stimulus from environment never comes alone, which increases the complexity of plant experiences, challenges and capacities (Trewavas 2003, 2009). Thus, response to new, ambiguous and unknown stimulus using past experiences can enhance plant performance in natural context, being an advantageous trait of plant life.

5 Concluding Remarks

Although acclimation (or hardening) and learning phenomena both are based on phenotypic plasticity, they implicate in different strategies for surviving and, ultimately, for plant fitness. Under the point of view of the observer, when plants are subjected to continuous environmental stimulus, progressive physiological and morphological changes often take place in order to reconcile plant organization with the new conditions. However, such changes may consist in very different strategies, depending on the external changes (stimulating or constraining development), ranging from improvements in some traits (e.g. higher water use efficiency, more efficiency in nutrient use) to morphological/physiological constraints, such as lower vital metabolic functions as photosynthesis and reduced growth. On the other hand, without the cessation of the stimulus, and the "imprinting" of memories, which could affect the future behaviour of the plant when subjected to a supposed similar environmental situation or a new situation that involves similar pathways of response (cross-tolerance phenomena), learning abilities are not accessible for the external observer. Furthermore, somehow, the past experiences shall improve the efficiency of future plant responses; otherwise, no learning at all would take place (when the responses are essentially the same), and, eventually, new stressful situations could cause more extensive damages, decreasing surviving and/or plant fitness.

Another major aspect to access plant learning is the problem of individuality and the scale of observation. As discussed earlier, insofar plants are modular organisms, consisting of modules with some level of autonomy, under heterogeneous environmental conditions, different parts of the plant body can exhibit different behaviours, not necessarily, underpinning learning phenomena. Additionally, a practical problem concerns on what techniques can be used to access plant behaviour. Physiological methods (including molecular ones) are restricted to specific scales of observation, for instance, ranging from the expression of individual genes to the whole canopy carbon assimilation. As argued by Vítolo et al. (2012) and Bertolli et al. (2014), due to the implications of the hierarchical organization of plants (Souza et al. 2016a), there is no a single representative scale of observation.

Thus, how does learning can be accessible for an external observer? Taking the conditions in the definition proposed herein, plant traits representing the integration of the whole plant module responses, such as biomass and fitness, would be the more reasonable manner to do that reliably. Therefore, the studies in smaller scales (e.g. molecular and cellular), seeking for the understanding of the causes of changes in plant behaviour in upper scales, must be cautious to make inferences and generalizations (Souza et al. 2016a).

Regardless of the difficulties to access learning, it has a major and self-evident role in plants' life. According to Trewavas (2003, 2005), the necessity for learning rather than rote behaviour relies on the enormous possibilities of environmental factor combinations creating an endless possible "worlds" for each individual plant. Thus, reactive preprogrammed responses are very unlikely to account for all sorts of plant demands in the wild. Accordingly, in order for learning process to occur, cues and signals of the aspects of the environment that are relevant to the learned adaptive response must be available to the organism (*environment invites behaviour*).

However, in some cases, the specific cues required to enable the organism to learn appropriately to their selective environment will not be clearly available (Brown 2013). This matches the concept of Sterelny's epistemic differences between two environmental categories (Sterelny 2003): "transparent" environments (where functional features correspond to reliable perceptual cues) and, more common in the wild, "opaque/translucent" environments (features do not correspond neatly to reliable cues). Thus, in opaque environments, learning may be a more costly process.

Brown (2013) draws attention to the fact the adaptive value of learning to individuals (and pop) is also sensitive to the cost-benefit structure of the world, insofar as plasticity has an intrinsic cost for plants (Kleunen and Fisher 2005). DeWitt et al. (1998) distinguish some potential costs of phenotypic plasticity: maintenance costs, relative to energetic costs of the sensory and regulatory mechanisms of PF; production costs associated to production of new inducible structures; information acquisition costs (like foraging exploitation of the environment); and genetic costs, for instance, when pleiotropic genes conferring adaptive plasticity on a trait also confer negative direct effects on other traits. Thus, learning process, taking place through phenotypic plasticity, is not always adaptive.

Finally, insofar learning can improve both surviving and plant fitness, there is also implication in the population level. Learning allows population to avoid the potential loss of genetic diversity that comes with directional selection, preserving the standing variation in populations (Brown 2013). Phenotypic plasticity (*underpinning learning process*) can reduce the lethality of the phenotypic variants within a population in the face of environmental changes, allowing higher stability (Souza and Lüttge 2015). This buffers population from future environmental changes by maintaining their capacity for rapid adaption in the future; ultimately, learning can be a source of lineage level of robustness and, in consequence, is a source of evolvability (Brown 2013).

The major problems in the world are the result of the difference between how nature works and the way people think. Gregory Bateson

References

Abramson CI, Chicas-Mosier AM (2016) Learning in plants: lessons from *Mimosa pudica*. Front Psychol 7:417

Affifi R (2013) Learning plants: semiosis between the parts and the whole. Biosemiotics 6:547–559

Agrawal AA (2002) Maternal effects associated with herbivory: mechanisms and consequences of transgenerational induced plant resistance. Ecology 83:3408–3415

Alpi A, Amrhein N, Bertl A, Blatt MR, Blumwald E, Cervone F, Dainty J, De Michelis MI, Epstein E, Galston AW, Goldsmith MH, Hawes C, Hell R, Hetherington A, Hofte H, Juergens G, Leaver CJ, Moroni A, Murphy A, Oparka K, Perata P, Quader H, Rausch T, Ritzenthaler C, Rivetta A, Robinson DG, Sanders D, Scheres B, Schumacher K, Sentenac H, Slayman CL, Soave C, Somerville C, Taiz L, Thiel G, Wagner R (2007) Plant neurobiology: no brain, no gain? Trends Plant Sci 12:135–136

Artetxe U, García-Plazaola JI, Hernández A, Becerril JM (2002) Low light grown duckweed plants are more protected against the toxicity induced by Zn and Cd. Plant Physiol Biochem 40:859–863

Atkinson NJ, Urwin PE (2012) The interaction of plant biotic and abiotic stresses: from genes to the field. J Exp Bot 63:3523–3544

Azooz MM, Ahmad P (eds) (2016) Plant-environment interaction. Wiley Blackwell, Hoboken

Bak P (1996) How nature works. The science of self-organized criticality. Springer, New York

Balmer A, Pastor V, Gamir J, Flors V, Mauch-Mani B (2015) The 'prime-ome': towards a holistic approach to priming. Trends Plant Sci 20:443–452

Baluška F (ed) (2013) Long-distance systemic signalling and communication in plants. Springer, Berlin

Baluška F, Mancuso S (2007) Plant neurobiology as paradigm shift not only in plant sciences. Plant Signal Behav 2:205–207

Baluška F, Mancuso S, Volkmann D (eds) (2006) Communication in plants: neuronal aspects of plant life. Springer, Berlin

Barlow P (2008) Reflections on 'plant neurobiology'. Biosystems 92:132–147

Barlow PW (2010) Plant roots: autopoietic and cognitive constructions. Plant Root 4:40–52

Bateson G (1972) Steps to an ecology of mind. Jason Aronson Inc., London

Beggs JM, Plenz D (2003) Neuronal avalanches in neocortical circuits. J Neurosci 23:11167–11177

Beggs JM, Plenz D (2004) Neuronal avalanches are diverse and precise activity patterns that are stable for many hours in cortical slice cultures. J Neurosci 24:5216–5229

Bertolli SC, Souza GM (2013) The level of environmental noise affects the physiological performance of *Glycine max* under water deficit. Theor Exp Plant Physiol 25:36–45

Bertolli SC, Mazzafera P, Souza GM (2014) Why is it so difficult to identify a single indicator of water stress in plants? A proposal for a multivariate analysis to assess emergent properties. Plant Biol 16:578–585

Bolker JA (2000) Modularity in development and why it matters to evo-devo. Am Zool 4:770–776

Brenner ED, Stahlberg R, Mancuso S, Vivanco J, Baluska F, Volkenburgh EV (2006) Plant neurobiology: an integrated view of plant signalling. Trends Plant Sci 11:413–419

Brown RL (2013) Learning, evolvability and exploratory behaviour: extending the evolutionary reach of learning. Biol Philos 28:933–955

Calvo P (2016) The philosophy of plant neurobiology: a manifesto. Synthese 193:1323–1343

Camazine S, Deneubourg J-L, Franks NR, Sneyd J, Theraulaz G, Bonabeau E (2001) Self-organization in biological systems. Princeton University Press, Princeton, NJ

Capiati DA, País SM, Téllez-Iñón MT (2006) Wounding increases salt tolerance in tomato plants: evidence on the participation of calmodulin-like activities in cross-tolerance signalling. J Exp Bot 57:2391–2400

Cayuela E, Perez-Alfocea K, Caro M, Bolarin MC (1996) Priming of seeds with NaCl induces physiological changes in tomato plants grown under salt stress. Physiol Plant 96:231–236

Cazalis R, Carletti T, Cottam R (2017) The living organism: strengthening the basis. Biosystems 158:10–16

Choh Y, Takabayashi J (2006) Herbivore-induced extrafloral nectar production in lima bean plants enhanced by previous exposure to volatiles from infested conspecifics. J Chem Ecol 32:2073–2077

Choi W, Hilleary R, Swanson SJ, Kim S, Gilroy S (2016) Rapid, long-distance electrical and calcium signalling in plants. Annu Rev Plant Biol 67:287–307

Clarke E (2012) Plant individuality: a solution to the demographer's dilemma. Biol Philos 27:321–361

Conrath U, Beckers GJ, Flors V, García-Agustín P, Jakab G, Mauch F, Newman MA, Pieterse CM, Poinssot B, Pozo MJ, Pugin A, Schaffrath U, Ton J, Wendehenne D, Zimmerli L, Mauch-Mani B (2006) Priming: getting ready for battle. Mol Plant-Microbe Interact 19:1062–1071

Crisp PA, Ganguly D, Eichten SE, Borevitz JO, Pogson BJ (2016) Reconsidering plant memory: intersections between stress recovery, RNA turnover, and epigenetics. Sci Adv 2:e1501340

Dat JF, Lopez-Delgado H, Foyer CH, Scott IM (1998) Parallel changes in H_2O_2 and catalase during thermotolerance induced by salicylic acid or heat acclimation in mustard seedlings. Plant Physiol 116:1351–1357

de Arcangelisa L, Herrmann HJ (2010) Learning as a phenomenon occurring in a critical state. Proc Natl Acad Sci U S A 107:3977–3981

de Kroon H, Huber H, Stuefer JF, van Groenendael JM (2005) A modular concept of phenotypic plasticity in plants. New Phytol 166:73–82

de Kroon H, Visser EJW, Huber H, Mommer L, Hutchings MJ (2009) A modular concept of plant foraging behaviour: the interplay between local responses and systemic control. Plant Cell Environ 32:704–712

De Loof A (2016) The cell's self-generated "electrome": the biophysical essence of the immaterial dimension of life? Commun Integr Biol 9:e1197446

Debono MW (2013a) Dynamic protoneural networks in plants: a new approach of spontaneous extracellular potential variations. Plant Signal Behav 8:e24207

Debono MW (2013b) Perceptive levels in plants: a transdisciplinary challenge in living organism's plasticity. Trans J Eng Sci 4:21–39

DeWitt TJ, Scheiner SM (2004) Phenotypic plasticity: functional and conceptual approaches. Oxford University Press, New York, NY

DeWitt TJ, Sih A, Wilson DS (1998) Costs and limits of phenotypic plasticity. Trends Ecol Evol 13:77–81

Engelberth J, Alborn HT, Schmelz EA, Tumlinson JH (2004) Airborne signals prime plants against insect herbivore attack. Proc Natl Acad Sci USA 101:1781–1785

Eurich CW, Herrmann JM, Ernst UA (2002) Finite-size effects of avalanche dynamics. Phys Rev E 66:066137

Feyerabend P (1975) Against method. NBL, London

Filippou P, Tanou G, Molassiotis A, Fotopoulos V (2012) Plant acclimation to environmental stress using priming agents. In: Tuteja N, Gill SS (eds) Plant acclimation to environmental stress. Berlin, NY, Springer Science & Business Media, pp 1–28

Firn R (2004) Plant intelligence: an alternative point of view. Ann Bot 93:345–351

Fromm J, Lautner S (2007) Electrical signals and their physiological significance in plants. Plant Cell Environ 30:249–257

Frost CJ, Heidi MA, Carlson JE, De Moraes CM, Mescher MC, Schultz JC (2007) Within-plant signalling via volatiles overcomes vascular constraints on systemic signalling and primes response against herbivores. Ecol Lett 10:490–498

Frost CJ, Mescher MC, Dervinis C, Davis JM, Carlson JE, De Moraes CM (2008) Priming defense genes and metabolites in hybrid poplar by the green leaf volatile cis-3-hexenyl acetate. New Phytol 180:722–734

Gagliano M (2015) In a green frame of mind: perspectives on the behavioural ecology and cognitive nature of plants. AoB Plants 7:plu075

Gagliano M, Renton M, Depczynski M, Mancuso S (2014) Experience teaches plants to learn faster and forget slower in environments where it matters. Oecologia 175:63–72

Gagliano M, Vyazovskiy VV, Borbeely AA, Grimonprez M, Depczynski M (2016) Learning by association in plants. Sci Rep 6:38427

Gallé A, Lautner S, Flexas J, Fromm J (2015) Environmental stimuli and physiological responses: the current view on electrical signalling. Env Exp Bot 114:15–21

Garzón FC (2007) The quest for cognition in plant neurobiology. Plant Signal Behav 2:208–211

Garzón FC, Keijzer F (2011) Plants: adaptive behaviour, root-brains, and minimal cognition. Adapt Behav 19:155–171

Gibson JJ (1966) The senses considered as perceptual systems. Houghton Mifflin, Boston

Gilroy S, Trewavas A (2001) Signal processing and transduction in plant cells: the end of the beginning? Nat Rev Mol Cell Biol 2:307–314

Goh CH, Gil Nam H, Shin Park Y (2003) Stress memory in plants: a negative regulation of stomatal response and transient induction of rd22 gene to light in abscisic acid-entrained Arabidopsis plants. Plant J 36:240–255

Gomila T, Calvo P (2008) Directions for an embodied cognitive science: toward an integrated approach. In: Calvo P, Gomila T (eds) Handbook of cognitive science: an embodied approach. Elsevier, San Diego, pp 1–26

Heil M, Kost C (2006) Priming of indirect defences. Ecol Lett 9:813–817

Heil M, Silva Bueno JC (2007) Within-plant signaling by volatiles leads to induction and priming of an indirect plant defense in nature. Proc Natl Acad Sci U S A 104:5467–5472

Heil M, Ton J (2008) Long-distance signalling in plant defence. Trends Plant Sci 13:264–272

Hilker M, Schwachtje J, Baier M, Balazadeh S, Bäurle I, Geiselhardt S, Hincha DK, Kunze R, Mueller-Roeber B, Rillig MC, Rolff J, Romeis T, Schmülling T, Steppuhn A, van Dongen J, Whitcomb SJ, Wurst S, Zuther E, Kopka J (2016) Priming and memory of stress responses in organisms lacking a nervous system. Biol Rev Camb Philos Soc 91:1118–1133

Hirao T, Okazawa A, Harada K, Kobayashi A, Muranaka T, Kirata K (2012) Green leaf volatiles enhance methyl jasmonate response in Arabidopsis. J Biosci Bioeng 114:540–545

Holeski LM (2007) Within and among generation phenotypic plasticity in trichrome density of *Mimulus guttatus*. J Evol Biol 20:2092–2100

Holeski LM, Jander G, Agrawal AA (2012) Transgenerational defense induction and epigenetic inheritance in plants. Trends Ecol Evol 27(11):618–626

Hossain MA, Fujita M (2013) Hydrogen peroxide priming stimulates drought tolerance in mustard (*Brassica juncea* L.) seedlings. Plant Gene Trait 4:109–123

Hossain MA, Mostofa MG, Fujita M (2013a) Cross protection by cold-shock to salinity and drought stress-induced oxidative stress in mustard (*Brassica campestris* L.) seedlings. Mol Plant Breed 4:50–70

Hossain MA, Mostofa MG, Fujita M (2013b) Heat-shock positively modulates oxidative protection of salt and drought-stressed mustard (*Brassica campestris* L.) seedlings. J Plant Sci Mol Breed 2:1–14

Hossain MA, Bhattacharjee S, Armin SM, Qian P, Xin W, Li HY, Burritt DJ, Fujita M, Tran LS (2015) Hydrogen peroxide priming modulates abiotic oxidative stress tolerance: insights from ROS detoxification and scavenging. Front Plant Sci 6:420

Hussain S, Khan F, Hussain HA, Nie L (2016) Physiological and biochemical mechanisms of seed priming-induced chilling tolerance in rice cultivars. Front Plant Sci 7:116

Hütt M-T, Lüttge U (2005) Network dynamics in plant biology: current progress in historical perspective. Prog Bot 66:277–310

Iqbal M, Ashraf M (2007) Seed preconditioning modulates growth, ionic relations, and photosynthetic capacity in adult plants of hexaploid wheat under salt stress. J Plant Nutr 30:381–396

Jakab G, Ton J, Flors V, Zimmerli L, Métraux JP, Mauch-Mani B (2005) Enhancing Arabidopsis salt and drought stress tolerance by chemical priming for its abscisic acid responses. Plant Physiol 139:267–274

Jenks MA, Hasegawa PM (2014) Plant abiotic stress. Wiley Blackwell, Hoboken

Jisha KC, Puthur JT (2016) Seed priming with beta-amino butyric acid improves abiotic stress tolerance in rice seedlings. Rice Sci 23:242–254

Jisha KC, Vijayakumari K, Puthur JT (2013) Seed priming for abiotic stress tolerance: an overview. Acta Physiol Plant 35:1381–1396

Jung HW, Tschaplinski TJ, Wang L, Glazebrook J, Greenberg JT (2009) Priming in systemic plant immunity RID D-4021-2009. Science 324:89–91

Kandel E, Dudai Y, Mayford M (2014) The molecular and systems biology of memory. Cell 157:163–186

Karban R (2015) Plant sensing and communication. The University of Chicago Press, Chicago

Kessler A, Halitschke R, Diezel C, Baldwin IT (2006) Priming of plant defense responses in nature by airborne signaling between *Artemisia tridentata* and *Nicotiana attenuata*. Oecologia 148:280–292

Kinouchi O, Copelli M (2006) Optimal dynamical range of excitable networks at criticality. Nat Phys 2:348–351

Knight H, Brandt S, Knight MR (1998) A history of stress alters drought calcium signalling pathways in Arabidopsis. Plant J 16:681–687

Kohler A, Schwindling S, Conrath U (2002) Benzothiadiazole-induced priming for potentiated responses to pathogen infection, wounding, and infiltration of water into leaves requires the NPR1/NIM1 gene in Arabidopsis. Plant Physiol 128:1046–1056

Kreps JA, Wu Y, Chang H-S, Zhu T, Wang X, Harper JF (2002) Transcriptome changes for Arabidopsis in response to salt, osmotic, and cold stress. Plant Physiol 130:2129–2141

Kron AP, Souza GM, Ribeiro RF (2008) Water deficiency at different developmental stages of glycine max can improve drought tolerance. Bragantia 67:693–699

Larcher W (1995) Physiological plan ecology, 3rd edn. Springer, Berlin

Li T, Holopainen JK, Kokko H, Tervahauta AI, Blande JD (2012) Herbivore-induced aspen volatiles temporally regulate two different indirect defences in neighbouring plants. Funct Ecol 26:1176–1185

Lucas M, Laplaze L, Bennett MJ (2011) Plant systems biology: network matters. Plant Cell Environ 34:535–553

Lüttge U (2008) Physiological ecology of tropical plants, 2nd edn. Springer, Berlin

Lüttge U (2012) Modularity and emergence: biology's challenge in understanding life. Plant Biol 14:865–871

Masi E, Ciszak M, Stefano G, Renna L, Azzarello E, Pandolfi C, Mugnai S, Baluska F, Arecchi FT, Mancuso S (2009) Spatiotemporal dynamics of the electrical network activity in the root apex. Proc Natl Acad Sci U S A 106:4048–4053

Mateo A, Mühlenbock P, Rustérucci C, Chang CC, Miszalski Z, Karpinska B, Parker JE, Mullineaux PM, Karpinski S (2004) LESION SIMULATING DISEASE 1 is required for acclimation to conditions that promote excess excitation energy. Plant Physiol 136:2818–2830

Maturana HR, Varela FJ (1980) Autopoiesis and cognition: the realization of the leaving. D. Reidel Publishing Company, London

Matyssek R, Lüttge U (2013) Gaia: the Planet Holobiont. Nova Acta Leopold 114:325–344

McCormick AC, Unsicker SB, Gershenzon J (2012) The specificity of herbivore-induced plant volatiles in attracting herbivore enemies. Trends Plant Sci 17:303–310

Merilo E, Jõesaar I, Brosché M, Kollist H (2014) To open or to close: species-specific stomatal responses to simultaneously applied opposing environmental factors. New Phytol 202:499–508

Miller JG (1978) Living systems. McGraw-Hill Publishing Co., New York

Mitchell M (2009) Complexity: a guided tour. Oxford University Press, New York

Mittler R (2006) Abiotic stress, the field environment and stress combination. Trends Plant Sci 11:15–19

Molinier J, Ries G, Zipfel C, Hohn B (2006) Transgenerational memory of stress in plants. Nature 442:1046–1049

Møller AP, Swaddle JP (1997) Asymmetry, developmental stability and evolution. Oxford University Press, Oxford

Mott KA, Buckley TN (2000) Patchy stomatal conductance emergent collective behaviour. Trends Plant Sci 5:258–262

Muroi A, Ramadan A, Nishihara M, Yamamoto M, Ozawa R, Takabayashi J, Arimura G (2011) The composite effect of transgenic plant volatiles for acquired immunity to herbivory caused by inter-plant communications. PLoS One 6:e24594

Nicolis G, Prigogine I (1989) Exploring complexity: an introduction. WH Freeman, New York

Olfati-Saber R, Fax JA, Murray RM (2007) Consensus and cooperation in networked multi-agent systems. Proc IEEE 95:215–233

Paparella S, Araújo SS, Rossi G, Wijayasinghe M, Carbonera D, Balestrazzi A (2015) Seed priming: state of the art and new perspectives. Plant Cell Rep 34:1281–1293

Pastori GM, Foyer CH (2002) Common components, networks, and pathways of cross tolerance to stress. The central role of "redox" and abscisic acid-mediated controls. Plant Physiol 129:460–468

Peng J, van Loon JJA, Zheng S, Dicke M (2011) Herbivore-induced volatiles of cabbage (*Brassica oleracea*) prime defense responses in neighboring intact plants. Plant Biol 13:276–284

Peñuelas J, Filella I, Zhang X, Llorens L, Ogaya R, Lloret R, Comas P, Estiarte M, Terradas J (2004) Complex spatiotemporal phenological shifts as a response to rainfall changes. New Phytol 161:837–846

Pozo MJ, Van Der Ent S, Van Loon LC, Pieterse CMJ (2008) Transcription factor MYC2 is involved in priming for enhanced defence during rhizobacteria-induced systemic resistance in Arabidopsis thaliana RID A-9326-2011. New Phytol 180:511–523

Prasch CM, Sonnewald U (2015) Signalling events in plants: stress factors in combination change the picture. Env Exp Bot 114:4–14

Rasmann S, De Vos M, Casteel CL, Tian D, Halitschke R, Sun JY, Agrawal AA, Felton GW, Jander G (2012) Herbivory in the previous generation primes plants for enhanced insect resistance. Plant Physiol 158:854–863

Richardson MJ, Shockley K, Fajen BR, Riley MA, Turvey MT (2008) Ecological psychology: six principles for an embodied–embedded approach to behaviour. In: Calvo P, Gomila T (eds) Handbook of cognitive science: an embodied approach. Elsevier, San Diego, pp 161–188

Rodriguez-Saona CR, Rodriguez-Saona LE, Frost CJ (2009) Herbivore-induced volatiles in the perennial shrub, *Vaccinium corymbosum*, and their role in inter-branch signaling. J Chem Ecol 35:163–175

Saraiva GFR, Ferreira AS, Souza GM (2017) Osmotic stress decreases complexity underlying the electrophysiological dynamic in soybean. Plant Biol 19:702–708

Schneider ED, Kay JJ (1994) Life as a manifestation of the second law of thermodynamics. Math Comput Model 19:25–48

Schroeder M (1991) Fractals, Chaos, Power Laws, Minutes From an Infinite Paradise. New York, WH Freeman and Company

Sheth BP, Thaker VS (2014) Plant systems biology: insights, advances and challenges. Planta 240:33–54

Shew WL, Yang H, Petermann T, Roy R, Plenz D (2009) Neuronal avalanches imply maximum dynamic range in cortical networks at criticality. J Neurosci 29:15595–15600

Slaughter A, Daniel X, Flors V, Luna E, Hohn E, Mauch-Mani B (2012) Descendants of primed Arabidopsis plants exhibit resistance to biotic stress. Plant Physiol 158:835–843

Souza GM, Lüttge U (2015) Stability as a phenomenon emergent from plasticity—complexity—diversity in eco-physiology. Prog Bot 76:211–239

Souza GM, de Oliveira RF, Cardoso VJM (2004) Temporal dynamics of stomatal conductance of plants under water deficit: can homeostasis be improved by more complex dynamics. Arq Biol Tecnol Curitiba 47:423–431

Souza GM, Pincus SM, Monteiro JAF (2005) The complexity-stability hypothesis in plant gas exchange under water deficit. Braz J Plant Physiol 17:363–373

Souza GM, Ribeiro RV, Prado CHBS, Damineli DSC, Sato AM, Oliveira MS (2009) Using network connectance and autonomy analyses to uncover patterns of photosynthetic responses in tropical woody species. Ecol Complex 6:15–26

Souza GM, Bertolli SC, Lüttge U (2016a) Hierarchy and information in a system approach to plant biology: explaining the irreducibility in plant ecophysiology. Progr Bot 77:167–186

Souza GM, Prado CHBA, Ribeiro RV, Barbosa JPRAD, Gonçalves AN, Habermann G (2016b) Toward a systemic plant physiology. Theor Exp Plant Physiol 28:341–346

Souza GM, Ferreira AS, Saraiva GFR, Toledo GRA (2017) Plant "electrome" can be pushed toward a self-organized critical state by external cues: evidences from a study with soybean seedlings subject to different environmental conditions. Plant Signal Behav 12:e1290040

Sterelny K (2003) Thought in a hostile world: the evolution of human cognition. Wiley-Blackwell, Oxford

Struik PC, Yin X, Meinke H (2008) Plant neurobiology and green plant intelligence: science, metaphors and nonsense. J Sci Food Agric 88:363–370

Sweetlove LJ, Fernie AR (2005) Regulation of metabolic networks: understanding metabolic complexity in the systems biology era. New Phytol 168:9–24

Thellier M, Lüttge U (2012) Plant memory: a tentative model. Plant Biol 15:1–12

Ton J, D'Alessandro M, Jourdie V, Jakab G, Karlen D, Held M, Mauch-Mani B, Turlings TC (2007) Priming by airborne signals boosts direct and indirect resistance in maize. Plant J 49:16–26

Trewavas A (2003) Aspects of plant intelligence. Ann Bot 92:1–20

Trewavas A (2005) Green plants as intelligent organisms. Trends Plant Sci 10:413–419

Trewavas A (2007) Response to Alpi et al.: Plant neurobiology – all metaphors have value. Trends Plant Sci 12:231–233

Trewavas A (2009) What is plant behaviour? Plant Cell Environ 32:606–616

Trewavas A (2014) Plant behaviour and intelligence. Oxford University Press, Oxford

Turlings TCJ, Ton J (2006) Exploiting scents of distress: the prospect of manipulating herbivore-induced plant odours to enhance the control of agricultural pests. Curr Opin Plant Biol 9:421–427

Van Kleunen M, Fisher M (2005) Constraints on the evolution of adaptive phenotypic plasticity in plants. New Phytol 166:49–56

van Wees SCM, Van der Ent S, Pieterse CMJ (2008) Plant immune responses triggered by beneficial microbes. Curr Opin Plant Biol 11:443–448

Verhagen BW, Glazebrook J, Zhu T, Chang HS, van Loon LC, Pieterse CM (2004) The transcriptome of rhizobacteria-induced systemic resistance in Arabidopsis. Mol Plant Microb Interact 17:895–908

Verhoeven KJ, van Gurp TP (2012) Transgenerational effects of stress exposure on offspring phenotypes in apomictic dandelion. PLoS One 7:e38605

Vialet-Chabrand S, Matthews JSA, Simkin AJ, Raines CA, Lawson T (2017) Importance of fluctuations in light on plant photosynthetic acclimation. Plant Physiol 173(4):2163–2179 (in press)

Vítolo HF, Souza GM, Silveira JAG (2012) Cross-scale multivariate analysis of physiological responses to high temperature in two tropical crops with C3 and C4 metabolism. Env Exp Bot 80:54–62

Von Bertalanffy L (1968) General system theory. George Braziller, New York

Walter J, Nagy L, Heinb R, Rascher U, Beierkuhnleinb C, Willner E, Jentsch A (2011) Do plants remember drought? Hints towards a drought-memory in grasses. Environ Exp Bot 71:34–40

Watling JR, Robinson SA, Woodrow IE, Osmond CB (1997) Responses of rainforest understorey plants to excess light during sunflecks. Aust J Plant Physiol 24:17–25

Withagen R, Poel HJ, Araújo D, Pepping G-J (2012) Affordances can invite behavior: reconsidering the relationship between affordances and agency. New Ideas Psychol 30:250–258

Witzany G (2006) Plant communication from biosemiotic perspective. Plant Signal Behav 1:169–178

Xia JH, Saglio PH (1992) Lactic acid efflux as a mechanism of hypoxic acclimation of maize root tips to anoxia. Plant Physiol 100:40–48

Zheng SJ, Dicke M (2008) Ecological genomics of plant–insect interactions: from gene to community. Plant Physiol 146:812–817

Mycorrhizal Networks Facilitate Tree Communication, Learning, and Memory

Suzanne W. Simard

Abstract Mycorrhizal fungal networks linking the roots of trees in forests are increasingly recognized to facilitate inter-tree communication via resource, defense, and kin recognition signaling and thereby influence the sophisticated behavior of neighbors. These tree behaviors have cognitive qualities, including capabilities in perception, learning, and memory, and they influence plant traits indicative of fitness. Here, I present evidence that the topology of mycorrhizal networks is similar to neural networks, with scale-free patterns and small-world properties that are correlated with local and global efficiencies important in intelligence. Moreover, the multiple exploration strategies of interconnecting fungal species have parallels with crystallized and fluid intelligence that are important in memory-based learning. The biochemical signals that transmit between trees through the fungal linkages are thought to provide resource subsidies to receivers, particularly among regenerating seedlings, and some of these signals appear to have similarities with neurotransmitters. I provide examples of neighboring tree behavioral, learning, and memory responses facilitated by communication through mycorrhizal networks, including, respectively, (1) enhanced understory seedling survival, growth, nutrition, and mycorrhization, (2) increased defense chemistry and kin selection, and (3) collective memory-based interactions among trees, fungi, salmon, bears, and people that enhance the health of the whole forest ecosystem. Viewing this evidence through the lens of tree cognition, microbiome collaborations, and forest intelligence may contribute to a more holistic approach to studying ecosystems and a greater human empathy and caring for the health of our forests.

S. W. Simard (✉)
Department of Forest and Conservation Sciences, Faculty of Forestry, University of British Columbia, Vancouver, BC, Canada
e-mail: suzanne.simard@ubc.ca

1 Introduction

Today, plants are commonly recognized as microbiomes—where villages of collaborative microbes live in and on their roots, stems, and leaves, forming interaction networks (Faust and Raes 2012; van der Heijden and Hartmann 2016). These microbial networks of fungi, bacteria, archaea, viruses, protists, and algae, as well as nematodes, arthropods, and protozoa (together comprising a soil food web), work together with the plants in complex adaptive systems to drive nature's biogeochemical cycles and influence every aspect of ecosystem structure and function (Ingham et al. 1985; Levin 2005). The interaction networks are highly coevolved and finely attuned, such that the loss of subjects from this village, particularly keystone species, could trigger system shifts to alternative stable states (Scheffer et al. 2001). The interaction among microbiomes and plants is so fundamental to life on earth that it is credited with the chemical weathering of rock and migration of ancient plants from the ocean to land about 360 Mya and the subsequent coevolution of highly specialized gymnosperm and angiosperm trees and ultimately humans (Margulis 1981; Humphreys et al. 2010; Archibald 2011). The rhizosphere (root-soil interface) microbiome is particularly diverse and active, with plants investing 10–90% of their photosynthate belowground to fuel rhizosphere processes involved in the carbon, nutrient, and water cycles, with the smallest proportion allocated belowground in the tropical forest biome and the largest in the grassland and tundra biomes (Poorter et al. 2012). The plant and microbial species that inhabit this rich zone have coevolved sophisticated communication systems to facilitate their multifarious interactions, where information is exchanged among organisms both within and among kingdoms (Baluška and Mancuso 2013).

The microbiome of the rhizosphere includes mycorrhizas (literally "fungus-roots")—generally mutualistic and obligate symbioses between root-inhabiting fungi and plants, involving 95% of plant families (Trappe 1987). Plants benefit by engaging with the fungus because it is energetically less expensive to invest in hyphal growth than root growth to acquire soil nutrients since complex compounds like cellulose and lignin are not required, and the fungal hyphae grow faster, have smaller diameters for accessing tight soil pores, and branch more profusely. The development of the mycorrhiza involves coevolved communication between the highly active plant root apex (Darwin's "root-brain"; see Baluška et al. 2010; Baluška and Mancuso 2013; more below) and the fungal symbiont, involving bidirectional elicitor signal molecules such as auxins, signal perception, signal transduction, and defense gene activation (Garcia-Garrido and Ocampo 2002). Once the mycorrhizal association is developed, the mycorrhizal fungus exchanges nutrients it forages with its extramatrical mycelium from the soil for photosynthate fixed by the plant. To meet plant nutrient and water demands, the roots and fungal hyphae must explore large volumes of soil to acquire the limiting and patchy resources (Smith and Read 2008), involving cognitive behaviors such as decision-making, search and escape movements, and neighbor recognition (Baluška et al. 2010; Heaton et al. 2012). Without their mycorrhizal fungal partners, the vast majority of plants could not acquire enough soil nutrients and water to grow, survive, and reproduce.

Mycorrhizal fungi can link the roots of different plant hosts, forming mycorrhizal networks (Molina and Horton 2015). Mycorrhizal networks are considered common across biomes because most mycorrhizal symbioses are generic, where a plant species associates with a diverse suite of fungal species or, conversely, a fungal species colonizes many plant species. Some of the associations are highly specialized, however, where some plant and fungal species only associate with a single partner species, with the potential to form exclusive, conspecific networks (Molina et al. 1992). In forests, heterospecific or conspecific networks of ectomycorrhizal fungi (EMF) form among gymnosperm and some angiosperm trees as well as woody shrubs in temperate and boreal forest biomes, whereas networks of the arbuscular mycorrhizal fungi (AMF) form mainly among angiosperm trees along with many herbs and grasses in the tropical forest biome, as well as some conifers (e.g., Cupressaceae and Aceraceae) in temperate forests (Smith and Read 2008). Ectomycorrhizal fungi occur predominantly in the *Basidiomycota* and *Ascomycota* phyla and are characterized by a fungal sheath around the root tip, a Hartig net enveloping the plant host root cell wall, and extramatrical mycelia, whereas the endomycorrhizal AMF occur predominantly in the *Glomeromycota* phylum, and these form arbuscules and sometimes vesicles inside the plant host root cells. Some exceptional plant families and genera are capable of forming viable symbioses with EMF and AMF simultaneously (e.g., *Salicaceae*, *Eucalyptus*) and serve as key hubs linking together ectomycorrhizal and arbuscular mycorrhizal networks (Molina and Horton 2015). Other endomycorrhizal classes include ericoid mycorrhizal fungi on autotrophic plant species in the Ericaceae family, arbutoid mycorrhizas on autotrophic plants in the Ericaceae subfamily Arbutoideae, monotropoid mycorrhizas on heterotrophic and mixotrophic plants in the subfamily Monotropoideae of the Ericaceae, as well as several genera in the Orchidaceae, and orchid mycorrhizas on heterotrophic orchids.

Plants, including trees, are increasingly understood to have cognitive capacity for perceiving, processing, and communicating with other plants, organisms, and the environment and to remember and use this information to learn, adjust their behaviors, and adapt accordingly (Gagliano 2014). In other words, plants are increasingly recognized as having agency that leads to decisions and actions, characteristics of intelligence usually only ascribed to humans or perhaps animals (Brenner et al. 2006). This recognition, that plants have agency and actions, in their capacity to perceive, communicate, remember, learn, and behave, could be transformative for how humans perceive, empathize with, and care for trees and the environment.

Trees are known to perceive and communicate with each other and other plants through root pathways (Baluška et al. 2010; Bierdrzycki et al. 2010) or using airborne signals (Heil and Karban 2009). They can also recognize the identity of neighboring plants and whether they are genetically related through root exudates (Bierdrzycki et al. 2010) or mycorrhizas (Pickles et al. 2016). Baluška and Mancuso (2013) propose that within- and between-plant communication is accomplished primarily via signal transport within and between roots, where compounds such as auxins serve as neurotransmitters across synapses at cell cross-walls within roots, across synapses between the apices of different plant roots, or between plant roots and symbiotic microbes and fungi in the rhizosphere. Because all trees are mycorrhizal in nature and mycorrhizal networks are considered ubiquitous in forests (Horton 2015), I propose that most belowground communication between trees in nature is mediated by mycorrhizas and that mycorrhizal networks are intimately involved in tree cognition. This follows closely on Baluška and Mancuso's (2013) recognition that communication between plants, and the involvement of cell-to-cell synapses and neurotransmitter-like compounds, has coevolved with microorganisms. Yet, much of the historic research on plant communication and cognition has been conducted on non-mycorrhizal plants grown in the lab or has not reported on the role of mycorrhizal fungi. The present review seeks to help set the stage for more holistic examinations of various aspects of plant cognition by involving their mycorrhizas in nature.

This chapter aims to review the fundamental role of mycorrhizal fungal networks in communication between trees and the functional, ecological, and evolutionary significance of this communication to forest communities in nature. I will review existing experimental evidence for cognition among trees facilitated by mycorrhizas, showcasing examples from the research in my lab. It is my hope that this might lead to an integrated approach to studying plant cognition in natural ecosystems that includes plant microbiomes.

2 Evidence for Tree Cognition Facilitated by Mycorrhizas

Cognition of plants, and the complex adaptive behaviors it triggers for enhanced fitness, requires perception, agency, and action (Gagliano 2014). While cognition and intelligence are usually considered exclusively the domain of humans and perhaps animals due to the existence of their central nervous systems, scientists in the field of plant cognition have effectively provided scientific evidence and argued for neuronal aspects in plants. This includes the existence of plant cell cross-walls, plasmodesmata, and synapses at root apices, analogous to neural synapses; signaling molecules that cross these synapses and transmit information via calcium-regulated exocytosis and vesicle recycling to neighboring cells, similar to neurotransmitters; and action potentials that rapidly transmit electrochemical signals to control plant physiology and behaviors, similar to a central nervous system (Baluška et al. 2005). Baluška et al. (2005) extend this concept to include cells of microbes in symbiosis with plants such as fungi and bacteria, where adjacent or interfacing plasma membranes form immunological synapses with plant cell membranes and molecules cross from plant–cell to microbe–cell, as in the trade of carbon and nutrient molecules across the plant–fungal membranes in mycorrhizas. Trees and plants use this neuronal physiology to then perceive the affordances of their environment (Gagliano 2014) through multiple sensory organs, including their leaves, roots, and microbiome (Karban et al. 2014; Bierdrzycki et al. 2010; van der Heijden and Hartmann 2016).

Yet, the absence of a brain, and its vast system of neurons, neurotransmitters, and action potentials organized as nodes and links in a complex modular neural network, now considered fundamental to neural plasticity, flexibility, and hence intelligence, questions the position that plant cognition is sophisticated and intelligent (e.g., good at making decisions, planning, organizing behaviors, solving problems, etc.) (Brenner et al. 2006; Gagliano 2014; Barbey 2017). Charles and Francis Darwin controversially proposed, with their "root-brain" hypothesis, that the root apex, located between the apical meristem and elongation zone of a root tip, acts like a brain-like organ that controls plant behavior, as with animals (Darwin 1880). Baluška et al. (2009) provide support for this hypothesis with the existence of "animal-like sensory-motoric circuits which allow adaptive behavior" such as root crawling and plant tropisms. However, in my view, the "root-brain" hypothesis cannot on its own adequately explain the sophisticated plant behaviors we observe in roots because, by nature of their energy-expensive constitution of cellulose, they lack the degree of flexibility needed to rapidly develop new transient pathways for tacking unique problems. Moreover, the "root-brain" hypothesis does not adequately fit with the new network neuroscience showing that general intelligence (g) arises from the existence of both "crystallized intelligence' (similar to memory) resulting from strong, well-worn overlapping pathways (or bonds) that access easy-to-reach network states, as well as "fluid intelligence" (similar to learning) resulting from weaker, more transient pathways and connections that access difficult-to-reach network states (Barbey 2017). To help complete the picture, I posit that when plants enter into symbioses with mycorrhizal fungi, this provides them with the necessary topology and energetics for sophisticated intelligence. Evidence for this follows.

3 Topology of Mycorrhizal Networks

Network topology refers to the arrangement of the various elements (nodes, links) of a communication network. In your brain, nodes and links could be neurons and axons; in a forest, they could be trees and interconnecting mycorrhizal fungal mycelia. Network topology determines how freely nodes interact with each other; how intense, frequent, or efficient their interactions are; and how vulnerable or resilient the network is to loss of specific nodes or groups of nodes (modules) (Bascompte 2009). Research in network neuroscience shows that general intelligence (g) is positively correlated with neural network architecture that is scale-free (the distribution of links per node follows a power-law, where there are a few highly connected nodes (i.e., hubs) and many weakly connected nodes) with small-world properties (cliquish, with frequent and strong links within cliques) (Barbey 2017). The scale-free network topology contrasts with random or regular networks, whose links are distributed more equally among nodes (Fig. 1). This architecture balances local with global efficiencies by having high local clustering (hubs, cliques, modules) and short path lengths (distance between distal nodes or clusters), allowing low-cost short-distance connections as well as shortcuts via hops and skips among distal nodes,

(a) Random network   (b) Scale-free network

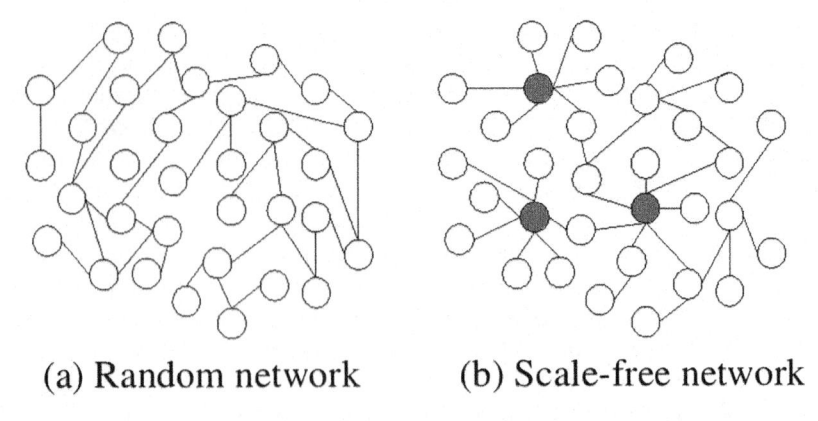

Fig. 1 (a) Random network and (b) scale-free network models. In forests, the circles (nodes) can represent trees, and the lines (links) can represent interconnecting mycorrhizal fungal genets. The random and scale-free networks differ by the pattern and accessibility of links. In random network models, each tree node is linked to a relatively small number of other nodes that are randomly distributed around the network. These networks can be easily traversed because there are few steps, or degrees of separation, between nodes. Random networks are more resilient to perturbations than scale-free networks. In scale-free models, the degree of links between nodes is variable, where some nodes, or hubs, are highly connected relative to the average (Barabási and Albert 1999). This model is more representative of living networks, such as mycorrhizal networks, which grow by accretion and have a dendritic form. Figure source: Wikipedia. Reproduced with permission from Simard (2012)

modules, or cliques that promote global information processing (Bray 2003). In your brain, you can think of modules or cliques as cortexes and lobes (e.g., frontal, temporal, parietal, etc.). In forests, modules could be clusters of trees, different species, or functional groups of species; or from a fungal perspective, they could be fungal species or functional groups such as exploration types. [Exploration types, including long distance, short distance, and contact, are distinguished based on the amount of emanating hyphae or the presence and differentiation of rhizomorphs and are considered important in accessing the diversity of soil substrates needed to supply trees with adequate nutrition (Agerer 2001)]. They are also important to modes of resource transmission through mycorrhizal networks (Teste et al. 2009; Hobbie and Agerer 2010). In either brains or mycorrhizal networks, modules are interconnected, albeit less frequently than within modules, by axons or mycorrhizal fungi.

In scale-free networks, this modular characteristic allows for specialized information processing while small-world properties allow for global and local efficiency and flexibility in memory-based learning. The presence of strongly connected modules and hubs supports linkage, nestedness, and short path lengths among nodes important in the mobilization of crystallized knowledge (i.e., memory) for learning. On the other hand, they also leave the network vulnerable to loss of key hubs (e.g., local injury to a brain lobe, high-grade logging, or pathological selection of the largest trees in a forest). The presence of weakly linked nodes (e.g., frontoparietal and cingulo-opercular networks or patches of small regenerating trees in forest gaps), by contrast, supports globally efficient small-world topology, access to difficult-to-reach states, and rapid adaptive behavior in novel situations (i.e., rapid learning) (Barbey 2017). In your brain, weak linkages can develop through rapid growth of synaptic connections between neurons and the myelination of nerve fibers, and these are strengthened via pruning in response to environmental conditions, which represents learning (Craik and Bialystok 2006). In mycorrhizal networks, weak linkages develop rapidly via cell expansion at growing mycelia fronts, where fungal apical tip growth, branching, anastomosis, and colonization of new plants occurs; this is thought to be accomplished predominantly by contact or short-distance explorer ectomycorrhizal fungal species (Agerer 2006; Hobbie and Agerer 2010; Heaton et al. 2012). This mycelium is very active, dynamic, and adaptive to simultaneously grow, prune, and regress in response to the rapidly changing environment, as in learning. It can also develop, via pruning, strong links that involve complex chords, strands, or rhizomorphs, predominantly formed by long-distance explorer fungi (Boddy and Jones 2007). These rhizomorphs not only exploit nutrients over short distances but also grow over long distances to reach disparate resource patches or form connections with distant ectomycorrhizal plants or modules (Agerer 2006; Lilleskov et al. 2011). They are capable of rapid, efficient high-volume resource transfer (Agerer 2006). You can think of the long-distance exploration rhizomorphs as analogous to "crystallized intelligence" and the rapidly expanding mycelial front of short-distance and contact explorers as "fluid intelligence." According to Barbey (2017), neuroscience research shows that this kind of scale-free network topology provides the greatest flexibility and dynamics that are crucial to learning and intelligence. It also shows that neural networks have the flexibility to transition between topologies, for example, from scale-free toward regular topology that is associated with more specific cognitive abilities and or toward a random topology that is associated with broader, more general abilities. Recent research in forest ecosystems also shows that mycorrhizal networks can transition from scale-free to regular and back to scale-free topology with the harvesting and planting of trees and subsequent stand development (van Dorp 2016). This dynamic flexibility likely underlies the diverse intelligence present among humans and forests.

Mycorrhizal network topology in forest stands, where trees are modeled as nodes and interconnecting fungal hyphae as links, is strikingly similar to the topology of neural networks in our brains (Southworth et al. 2005; Lian et al. 2006; Beiler et al. 2010, 2015; Toju et al. 2014). In Beiler et al. (2010, 2015), we used multi-locus, microsatellite DNA markers to show that most trees in uneven-aged forests of Douglas-fir (*Pseudotsuga menziesii* var. *glauca*) were interconnected by mycorrhizal networks of two ectomycorrhizal fungi, *Rhizopogon vesiculosus* and *R. vinicolor*. These two sister species of *Rhizopogon* share narrow host specificity for Douglas-fir (Kretzer et al. 2003), and they dominate the diverse community of 65 ectomycorrhizal fungal species that occur in all stages of forest stand development (Twieg et al. 2007). The *Rhizopogon* species fruit in truffles belowground and have coralloid or tuberculate structures with fine, dark extramatrical hyphae capable of rapidly growing over short distances and forming highly differentiated rhizomorphs capable of transporting water and dissolved nutrients over long distances (Brownlee et al.

Fig. 2 *Rhizopogon vinicolor* in a dry interior Douglas-fir forest of British Columbia. (**a**) Fruiting body (truffle) on forest floor; (**b**) tubercle with outside rind removed tubercle; (**c**) tubercle, rhizomorphs, hyphae, and roots in mineral soil profile; and (**d**) mixed mycorrhizal hyphal network on surface of mineral soil (forest floor peeled off). Photos by Kevin J. Beiler and Hugues Massicotte. Reproduced with permission from Simard (2012)

1983; Molina 2013) (Fig. 2). We found that the short-distance hyphae and long-distance rhizomorphs of *R. vesiculosus* and *R. vinicolor* colonized trees of all sizes and ages, forming spatially continuous, complex networks linking together multiple trees in the forest (Beiler et al. 2012).

The *Rhizopogon*-Douglas-fir mycorrhizal network had a scale-free network structure with small-world properties, where a few large, old hub trees had the greatest number of fungal connections and were linked to many small, young trees that had fewer connections (Beiler et al. 2010, 2015) (Fig. 3). This architecture makes sense given that rooting density and extent, and hence density of mycorrhizal root tips and connection potential, is correlated with the size of a tree. In a 30×30m patch in one of the stands, a single hub tree was linked to 47 other trees and was estimated to be linked to at least 250 more trees had the larger stand been sampled. The veteran hub trees provided an extensive network into which almost all of the smaller and younger understory seedlings and saplings had established. The high clustering in the network suggested that the old hub trees provided network paths or hyperlinks that bridged

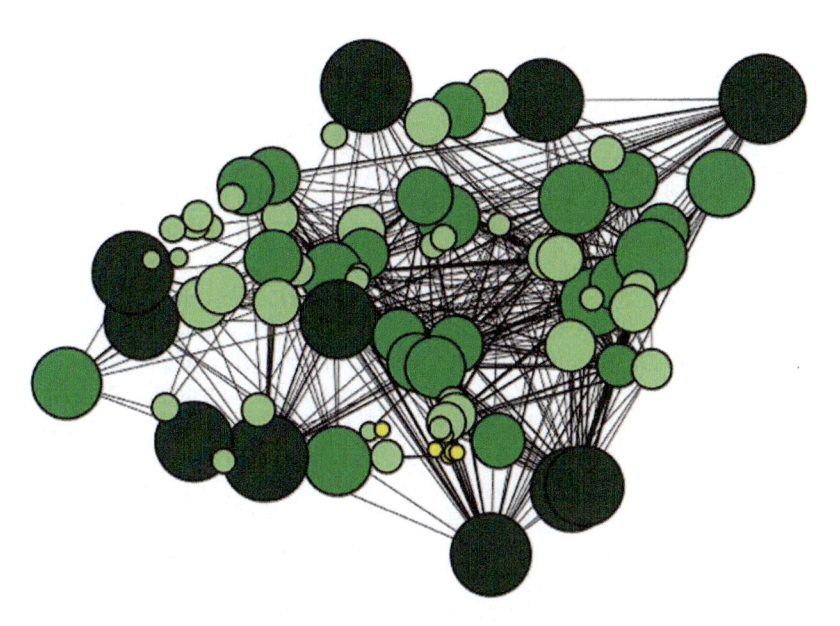

Fig. 3 Mycorrhizal network topology of Douglas-fir forest showing linkages between interior Douglas-fir trees via shared colonization by *Rhizopogon vesiculosus* and *R. vinicolor* genets. Circles represent tree nodes, sized according to the tree's diameter, and colored with four different shades of yellow or green that increase in darkness with increasing age class. Lines represent the Euclidean distances between trees that are linked. Line width increases with the number of links between tree pairs. Reproduced with permission from Beiler et al. (2010)

cliques (modules) of the densely interconnected younger trees. These pathways allowed the entire network to be easily traversed, which is a small-world property. The high density of fungal links within patches (modules) meant the patches were resilient to random disturbances but also vulnerable to attacks that target hubs. The *R. vinicolor* linkages were smaller and nested within the larger, denser *R. vesiculosus* network, forming a cliquish, nested "meta-network," and this nestedness increased network resilience. The network density and complexity is undoubtedly vastly more complex than we were able to describe given that we accounted for only two of the 65 ectomycorrhizal species in the forest, and we did not examine the arbuscular, ericoid, arbutoid, or orchidoid subnetworks associated with other tree and understory plant species that would have been nested within the *Rhizopogon* network.

4 Communication Through Mycorrhizal Networks

The scale-free, small-world network topology of the mycorrhizal network is designed for efficiency—for quickly shuttling signals through links among numerous trees, including between old hubs and young nodes, and for minimizing the costs of this information transmission while maximizing the impact on growth and adaptation of the network (Barbey 2017). Our numerous experiments have found that a multitude of signals—including nitrogen, carbon, water, defense molecules, and kin recognition information—transmit back-and-forth among Douglas-fir trees through ectomycorrhizal networks (for a review, see Simard et al. 2015). Phosphorus (Eason et al. 1991; Finlay and Read 1986; Perry et al. 1989), other defense signals (Song et al. 2010; Babikova et al. 2013), allelochemicals (Barto et al. 2011), nutrient analogues (Meding and Zasoski 2008; Gyuricza et al. 2010), and genetic material (Giovannetti et al. 2004, 2005) have also been shown to transmit through arbuscular networks or among different ectomycorrhizal plants in other studies. These compounds can be large or small and include fungal carbohydrates (e.g., trehalose, mannitol, arabitol, and erythritol, see below), amino acids (e.g., glutamine and glycine), lipids, N ions (NH_4^+ or NO_3^-), phosphates, and nuclei (Martin et al. 1986; Smith and Smith 1990; Bago et al. 2002; Giovannetti et al. 2005; Nehls et al. 2007). Phytohormones such as auxins and jasmonates, which are signals important in regulating the mycorrhizal symbiosis as well as plant phenotypic plasticity, have also been shown to converge in mycorrhizal hubs (Pozo et al. 2015). Most of these signals shuttle rapidly within and between the plants—within hours or a few days—and they are of sufficient magnitude to influence plant behaviors such as root foraging, nutrient acquisition, growth, or survival (Simard et al. 2012). Even faster and more efficient signaling between plants could occur via sound transmission (Gagliano 2012) through mycorrhizal networks, much like a conversation over the telephone, but this mode of communication has yet to be explored in mycorrhizal networks.

In my view, the transmission of signals or resources or molecules or sounds between plants through mycorrhizal fungal networks constitutes communication. The Latin root of the word communication is *communicat*, or *share*, and it is the transfer or sharing of information through a common system of signals that benefits both the sender and the receiver. As argued by Gagliano (2012), interplant signaling of information is now widely accepted among scientists as a form of communication between plants. Moreover, where signaling is communication, the signals that are communicated constitute language (Gagliano and Grimonprez 2015). Language can include spoken or written words, sounds, signals, or gestures used to communicate and is used by individuals, whether human, animal, or plant, to make sense of and survive in this world. In this sense, the chemistry or sound transmitted between plants is their language, and by analogy, the highly varied compounds or sound waves emitted constitute their vocabulary. This language has emerged from local repeated iterative interactions among plants, fungi, other organisms, and the environment, leading to increased fitness of the species by enhancing their adaptive capacity, learning capabilities, and ultimately coevolution (Gagliano and Grimonprez 2015). It allows plants to plastically adjust to environmental challenges, and this ability is enhanced by their associated microbiota.

Signals that are transmitted cell-to-cell and tree-to-tree through mycorrhizal networks can be considered analogous to neurotransmitters in biological neural networks. Some of the amino acids and phytohormones transmitted through mycorrhizal networks are structurally analogous to neurotransmitter transporters that are highly conserved in humans and animals (Wipf et al. 2002; Baluška et al. 2005). Auxin, for example, is structurally similar to serotonin (Pelagio-Flores et al. 2011; Baluška and Mancuso 2013). Glutamate is the most abundant excitatory neurotransmitter in the central nervous system and accounts for over 90% of synaptic transmissions in the human brain. Glycine is the most common inhibitory neurotransmitter and is highly active in the brain and spinal cord (Bowery and Smart 2006). Glutamine and glycine are also the primary amino acids through which nitrogen is transferred from EMF to their hosts (Martin et al. 1986; Taylor et al. 2004) and through which nitrogen and carbon are thought to transfer along source-sink gradients through mycorrhizal networks (Martin et al. 1986; Teste et al. 2009, 2010; Deslippe and Simard 2011; Simard et al. 2015; Deslippe et al. 2016).

These signals, the amino acids, hormones, and other compounds that constitute the language of plants, flow symplastically and apoplastically through the interlinking mycorrhizal hyphae and rhizomorphs of the network, crossing plant and fungal synapses and following source-sink gradients among tree and plant nodes (Simard et al. 2015). Leaf photosynthetic activity likely generates nitrogen and carbon source-sink gradients within donor plants that drive the transport of the amino acids into the mycorrhizal roots, followed by their transmission via mass flow through the interconnecting mycelium, and then up into the xylem of the linked receiver sink plants. Glutamine contains five C atoms for every two N atoms, and glycine contains two to one, reflecting the high-energy cost of N assimilation by plants (Martin et al. 1986; Taylor et al. 2004). When glutamine and glycine are delivered in high quantities from the mycelium to the plant (Yang et al. 2010), the plant would receive a significant C subsidy in addition to N, while the fungus would still receive its most limiting resource, C, from the plant. Teste et al. (2009) used dual isotope labeling with ^{13}C and ^{15}N to show that nitrogen-rich Douglas-fir saplings simultaneously transferred N and C to N-poor conspecific germinants through mycorrhizal networks and that this corresponded with greater 2-year receiver seedling survival. The relative amounts of N (0.0018%) and C (0.0063% of photo-assimilate) transferred had a stoichiometry of 2N:7C, which is similar to glutamine (2N:5C), alanine, and cysteine (2N:6C), but the transmitted compounds were never identified in that study (Teste et al. 2009). In the central nervous system, some of these amino acids (glutamate, cysteine) activate postsynaptic cells, whereas others (glycine, alanine) depress the activity of postsynaptic cells (Dehaene et al. 2003). In plants, these compounds are involved in basic metabolism, such as regulation of ion transport, modulation of stomatal opening, enzyme and protein synthesis, gene expression, etc. (Rai 2002).

Both the plant nodes and fungal links are involved in the regulation of interplant communication. Resources and signals transmit back-and-forth between plants through the fungal networks according to supply and demand or stress gradients in the plant communities, representing a complex underground trading system. This trading of information is like a conversation, where two or more plants and the fungi exchange information in a local setting. Patterns of transmission of C, N, water, and other information depends on source-sink gradients governed by factors such as physiological, nutrient or water status of the donor and receiver plants, stress gradients within the plant community, degree or dependency of these plants on mycorrhization, the fungal species involved in the network, or nutrient or water status of patchy soil environments. Numerous experiments have shown that differences in physiological source-sink strength or stress among plants (e.g., in photosynthetic rates, growth rates, nutrient content, age, defoliation by pathogens, insects or drought) influence transmission patterns (Simard and Durall 2004; Leake et al. 2004; Selosse et al. 2006; van der Heijden and Horton 2009; Song et al. 2010). Characteristics of fungal and associated microbial communities also play important roles (Finlay 1989; Rygiewicz and Anderson 1994; Lehto and Zwiazek 2011). The importance of the mycorrhizal fungi to interplant communication has been supported by experiment inoculations with different fungal species (Arnebrant et al. 1993; Ekblad and Huss-Danell 1995; Ek et al. 1996; He et al. 2004, 2005; Egerton-Warburton et al. 2007; Querejeta et al. 2012) and the use of mesh that allows certain fungal exploration types to join the network (Teste et al. 2009; Bingham and Simard 2012).

5 Plant Behavioral Responses, Learning, and Memory

Plant behavior responses and learning are actions or changes in plant morphology and physiology to environmental stimuli—these flow from the agents of cognition, which include their senses, mycorrhizal networks and signal transmission, as described above. These agents provide plants with sophisticated mechanisms for perceiving their environment, storing the information in their memory banks such as annual growth rings, seeds, or branching, rooting and network topologies, and using this information for memory-based learning that drives behaviors such as choice, decision-making, defense, and neighbor recognition. Communication between plants through mycorrhizal networks, for example, has been associated with behavioral shifts expressed as changes in rooting patterns, mycorrhizal network development, nutrient uptake, and defense enzyme production. These shifts have resulted in changes in survival, growth, and fitness of the sender and receiver plants. McNickle et al. (2009) define behavior as the expression of plant plasticity that is like a decision point, where each choice involves trade-offs that will affect fitness.

5.1 Behaviors

Plant behaviors that have been influenced by interplant signal transmission through mycorrhizal networks include, for example, changes in (1) plant morphology such as rooting depth, height growth, or mycorrhizal network patterns; (2) plant physiology such as photosynthetic rates, stomatal conductance, and nutrient uptake; and (3) plant fitness indicators such as germination, survival, and gene regulation of defense chemistry. These behavioral changes have been well described in previous reviews, including Selosse et al. (2006), Simard et al. (2012, 2013, 2015), Gorzelak et al. (2015), and Horton (2015). Here, I briefly summarize only those that have been associated with interplant transmission of carbon, nitrogen, and water through ectomycorrhizal networks.

Carbon fluxes through ectomycorrhizal networks are substantial and vary with the degree of heterotrophy of the plants; they supply up to 10% of autotrophic, up to 85% of partial mycoheterotrophic, and 100% of fully mycoheterotrophic plant carbon. This carbon supply has been associated with increased survival and growth of autotrophic plants and is essential for the survival of fully mycoheterotrophic plants. For example, in our Douglas-fir forests, seedling establishment success was significantly greater where seedlings had full access to the mycorrhizal networks of older Douglas-fir trees compared to where they did not (Teste et al. 2009; Bingham and Simard 2012). Access to the network of the old trees not only improved conspecific seedling survival, but seedlings were colonized by a more complex fungal community comprising multiple short- and long-distance exploration types. It is because of their ability to nurture the understory seedlings, many of them related (see below), that we have named the old hub trees "mother trees" (Simard 2012). In other plant and tree communities, network-mediated nitrogen fluxes from N_2-fixing plants have supplied up to 40% of receiver N to non-N_2-fixing plants, and this has been associated with increased plant productivity (e.g., He et al. 2003, 2005, 2009). Fluxes between non-N_2-fixing plants have supplied <5% of receiver N (e.g., He et al. 2006; Teste et al. 2009). Ectomycorrhizal networks also facilitate the hydraulic redistribution of soil or plant water following water potential gradients, including Douglas-fir forests, supplying up to 50% of plant water that is essential for plant survival and growth (see Simard et al. 2015).

5.2 Learning

Learning occurs when plant perceive their environment and use this information to modify their behavior for optimizing environmental resources to increase fitness. It can involve social learning, trial and error, cultural transmission, and epigenetics (Gagliano 2014). Here I provide two examples of mycorrhizal network-mediated social learning and epigenetics in plants involving kin recognition and defense signaling, which we previously described in Gorzelak et al. (2015).

Plants can recognize the degree of relatedness of neighboring plants through a process called kin recognition, change their behavior for optimally interacting with these neighbors, learn to respond to concurrent changes in the behavior of the neighbors, and in so doing increase fitness (Dudley and File 2008; Karban and Shiojiri 2009; Novoplansky 2009; Dudley et al. 2013; Asay 2013; Gorzelak 2017). Kin recognition has been shown in several experiments to be mediated by mycorrhizas or mycorrhizal networks (File et al. 2012a, b; Asay 2013; Pickles et al. 2016; Gorzelak 2017). For example, foliar nutrition in AMF *Ambrosia artemisiifolia* L. improved when it was integrated into a mycorrhizal network with related plants but not conspecific strangers (File et al. 2012a, b). Likewise, in pairs of EMF Douglas-fir seedlings grown in greenhouse conditions, growth attributes and foliar micronutrients were increased in kin compared with strangers grown with older conspecifics (Asay 2013). In both cases, mycorrhizal colonization was elevated in the related but not stranger neighbors, which led to increased growth and nutrition of both seedlings in the pair (File et al. 2012a, b; Asay 2013). These findings reveal that mycorrhizas and mycorrhizal networks can play an integral role in kin recognition and that learning from increased mycorrhization of kin enhanced the plant morphological and physiological responses. The exact mechanism by which kin recognition occurs, however, is unclear. Nevertheless, there is strong evidence that biochemical signals derived from mycorrhizas or roots are involved (Bierdrzycki et al. 2010; Semchenko et al. 2007). For example, Semchenko et al. (2007) showed that root exudates carried specific information about the genetic relatedness, population origin, and species identity of neighbors, and locally applied exudates triggered different root behavior responses of neighbors. This included increased root density, achieved through changes in morphology rather than biomass allocation, suggesting the plants learned from their neighbors to limit the energetic cost of their behavior. Because the overwhelming majority of plants are predominantly mycorrhizal in situ and because mycorrhizal networks are considered common in nature, any root exudates involved in kin recognition are likely to be filtered through mycorrhizal fungi and mycorrhizal networks. In a recent study using stable-isotope probing, we found that mycorrhizal networks transmitted more carbon from older donor Douglas-fir seedlings to the roots of younger kin receiver seedlings than to stranger receiver seedlings, suggesting a fitness advantage to genetically related neighbors (Fig. 4; Pickles et al. 2016). This may have been facilitated by the greater mycorrhizal colonization of kin than stranger seedlings (Asay 2013), creating a stronger sink in the mycorrhizal network, an effect also noted in the study by File et al. (2012a, b). Gorzelak (2017) later found that herbivory of Douglas-fir induced greater transfer of carbon through mycorrhizal networks to neighboring kin than stranger seedlings.

Defense signals travelling through mycorrhizal networks also result in rapid behavioral responses of recipient plants, and this is evident in sudden changes in foliar defense chemistry (Babikova et al. 2013; Song et al. 2015) and pest resistance (Song et al. 2010, 2014). For instance, broad beans (*Vicia faba*) responded to aphid attack by swiftly transferring defense signals via mycorrhizal networks to neighboring bean plants, which learned from this to produce aphid-repellent chemicals and aphid-predator attractants (Babikova et al. 2013). This learning represents a trophic cascade generated by pest infestation and signal propagation through the mycorrhizal network. In a different study, defoliation of Douglas-fir resulted in a simultaneous transfer of defense signals and carbon to neighboring healthy ponderosa pine through mycorrhizal networks. The ponderosa pine learned from these triggers to increase defense enzyme production and protect itself against the loss of healthy hosts (Song et al. 2015). In earlier studies, Song et al. (2010, 2014) showed that increases in mycorrhizal network-mediated enzyme production flowed from upregulation of defense genes and modification of gene expression, constituting an epigenetic effect. Responses to pest infestations can also lead to larger-scale generational changes in the behavior of plant-symbiont systems. Shifts in ectomycorrhizal community composition caused by a variety of factors, such as host mortality (e.g., pine beetle outbreaks; Kurz et al. 2008), can result in ecological memory effects that

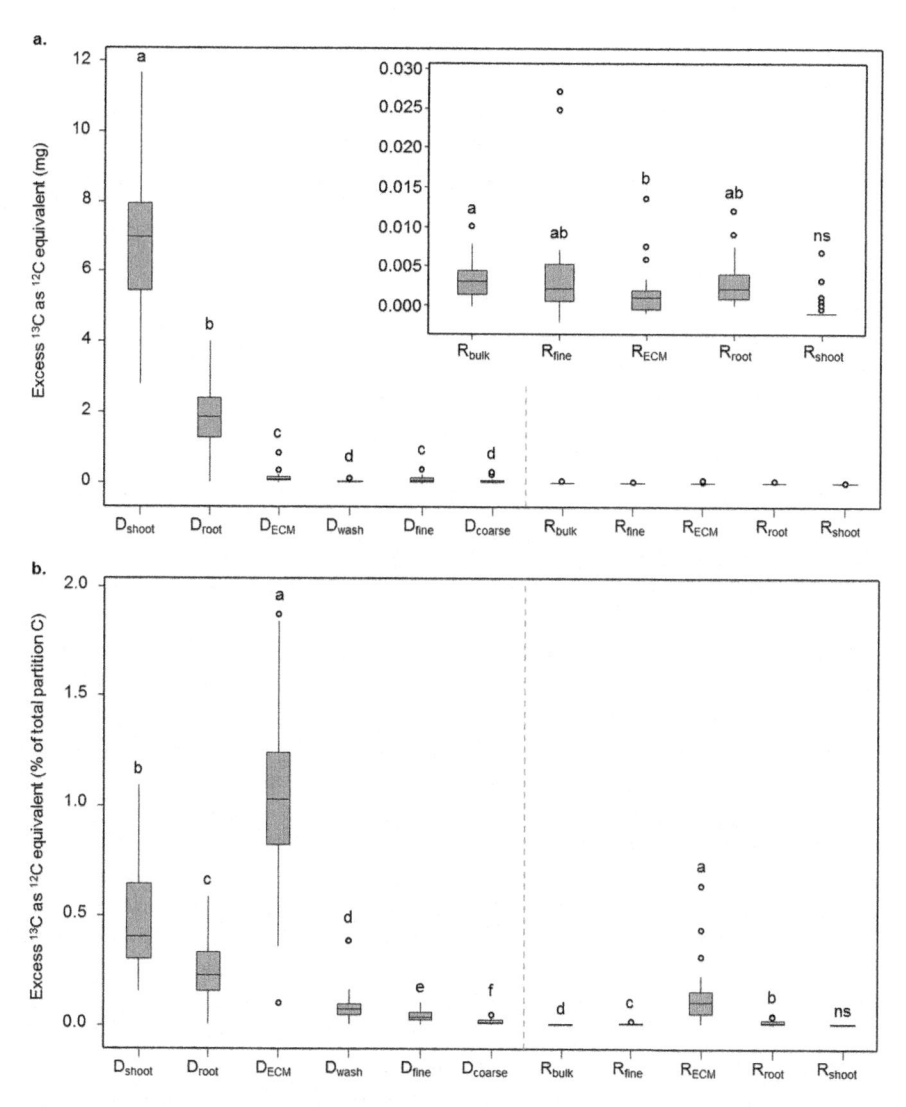

Fig. 4 Distribution of excess ^{12}C equivalent (mg) added to each tissue or soil partition in terms of (**a**) the total mass added and (**b**) the percentage of partition C comprised of enriched biomass, following labeling and a 6-day chase period for all donor (D) and kin recipient (R) partitions analyzed. Shoot, root, ECM, wash, and fine partitions represent their total biomass, whereas D_{coarse} and R_{bulk} soil partitions represent a subsampling of the entire soil environment. Inset shows rescaled recipient partitions for comparison. Letters indicate significant differences between partitions compared within D and R compartments. Reproduced with permission from Pickles et al. (2016)

impact future generations of the host species (Karst et al. 2015). For example, in areas of western North America dramatically impacted by the mountain pine beetle-induced dieback of lodgepole pine (*Pinus contorta*), EMF have declined significantly (Treu et al. 2014). Seedlings grown in soils from beetle-attacked pine stands learned from this decline and then expressed both reduced biomass and reduced production of monoterpenes compared with those grown in soil from undisturbed pine stands. This reveals a transgenerational cascade involving learning, memory, and epigenetics mediated by fungal symbionts (Karst et al. 2015).

5.3 Memory

Memory is a process by which organisms acquire, encode, store, and retrieve information. This information can then form the basis for experiential learning, where organisms modify their actions for improved fitness. One interesting example of memory-based learning is emerging from our new research in the salmon forests of the Pacific Coast.

We are studying now, how "mother trees"—the ancient cedars, spruces, and firs of the Pacific Coast—transmit nutrients via their massive fungal networks through the forest, feeding the entire ecosystem. Here is how we think this works. The salmon eggs hatch in the freshwater streams of the coastal forests, and then the fry, swim out to the sea, where the fish spend their adult lives feeding in the open ocean. Every spring and fall, the salmon return to their mother streams to breed and die, carrying with them nutrients from the ocean. The Aboriginal people of the Pacific Coast use the salmon for their livelihoods and have traditionally built tidal stone traps at the mouth of marine spawning rivers to passively catch the fish. Not only people but other predators and scavengers, including grizzles, wolves, and eagles, also feed on the carcasses. These creatures carry their catch up the riverbanks, settling to feast on the safe, warm, dry benches under the mother trees in the riparian forest. In so doing, they distribute the nutrients in the salmon carcasses and their feces and urine. The bears eat the innards in safety, leaving the carcasses to decay and the nutrients to seep into the soil. The mycorrhizal fungi associated with the roots of the trees and plants acquire the salmon nutrients from the soil and use them to supply 25–90% of the tree and plant nitrogen budgets. Once metabolized in the woody tissues of the trees, the salmon nutrients are stored in tree rings for centuries, providing a memory bank of historical salmon runs for as long as the tree is old. This process contributes to faster tree growth along salmon streams and underlies the great size and unparalleled productivity of these old forests. It has also been shown to shape the diversity and composition of vegetation, insect, and bird communities (Hocking and Reynolds 2011). This process of salmon nutrient uptake by the mycorrhizas, storage in tree rings, and retrieval of the information for tree growth, constitutes a memory embedded in the forest. We are examining now whether these salmon nutrient memories are transmitted from tree to tree and from plant to plant, through their fungal connections, deep into the forest. The spreading of the salmon memory, the telling of the story through their communication networks, allows the trees, fungi, bears, and salmon to collaboratively inform the productivity and health of the ecosystem. These luxurious forests in turn shade and nurture the salmon rivers, modulating the water temperature and transmitting nutrients to the ebb tides through seepage, thus forming a positive feedback loop that promotes the health and productivity of the fish. The parts of the trees—the bark and roots, made with salmon nitrogen—are harvested by the Aboriginal people to make clothing and art and tools, such as for the harvest of salmon. Mother trees play a crucial role in the closing of this circle. The health of the forest is thus tied to the health of the salmon, and it cycles back to the rivers, the oceans, and the people. The integrity of this circle of life depends on what the Aboriginals call reciprocity—the trade of mutual respect. This is an example of how people are sustainably embedded in this complex adaptive system.

This collective behavior, learning, and memory in the salmon forest may allow the community to solve cognitive problems that go beyond the capacity of a single organism, facilitating altruistic behaviors like kin recognition and more generally promoting cooperation for better ecosystem health.

6 Conclusions

This chapter has provided evidence that mycorrhizal networks are crucial agents in tree and plant perceptions of their neighbors and environment, in interplant communication of their strengths, needs and stresses, in the acquisition and storage of memories, and in memory-based learning and adaptive behaviors. The scale-free topology and small-world properties of mycorrhizal networks, along with similarity in transmitted signals to neurotransmitters in vertebrates, provide the necessary biological agency for intelligence in forests. The agency and conveyance of information through the mycorrhizal network provides manifold opportunities for trees to take action for interacting with their neighbors and adapting to the rapidly changing environment. Through sophisticated cognition that is facilitated by their microbiomes, trees and plants are more perceptive, intelligent, and in control of their destiny than humans have ever given them credit for. It is my hope that future research in plant cognition includes the crucial role of plant microbiomes, and mycorrhizal networks in particular.

Acknowledgments I thank my graduate students, postdoctoral fellows, and collaborators who contributed to this research on mycorrhizal networks over the years, including Amanda Asay, Jason Barker, Marcus Bingham, Camille Defrenne, Julie Deslippe, Dan Durall, Monika Gorzelak, Melanie Jones, Justine Karst, Allen Larocque, Deon Louw, Katie McMahen, Gabriel Orrego, Huamani Orrego, Julia Amerongen Maddison, Greg Pec, Leanne Philip, Brian Pickles, Teresa Ryan, Laura Super, Francois Teste, Brendan Tweig, and Matt Zustovic. This research was supported by a Natural Sciences and Engineering Research Council (NSERC) Discovery Grant to SWS.

References

Agerer R (2001) Exploration types of ectomycorrhizal mycelial systems: a proposal to classify mycorrhizal mycelial systems with respect to their ecologically important contact area with the substrate. Mycorrhiza 11:107–114

Agerer R (2006) Fungal relationships and structural identity of their ectomycorrhizae. Mycol Prog 5:67–107

Archibald JM (2011) Origin of eukaryotic cells: 40 years on. Symbiosis 54:69–86

Arnebrant K, Ek H, Finlay RD, Söderström B (1993) Nitrogen translocation between *Alnus glutinosa* (L.) Gaertn. seedlings inoculated with *Frankia* sp. and *Pinus contorta* Dougl. ex Loud seedlings connected by a common ectomycorrhizal mycelium. New Phytol 24:231–242

Asay AK (2013) Mycorrhizal facilitation of kin recognition in interior Douglas-fir (*Pseudotsuga menziesii* var. *glauca*). Master of Science thesis. University of British Columbia, Vancouver, Canada

Babikova Z, Gilbert L, Bruce TJA, Birkett M, Caulfield JC, Woodcock C, Pickett JA, Johnson D (2013) Underground signals carried through common mycelial networks warn neighbouring plants of aphid attack. Ecol Lett 16:835–843

Bago B, Zipfel W, Williams RM, Jun J, Arreola R, Lammers PJ, Pfeffer PE, Shachar-Hill Y (2002) Translocation and utilization of fungal storage lipid in the arbuscular mycorrhizal symbiosis. Plant Physiol 128:109–124

Baluška F, Mancuso S (2013) Microorganism and filamentous fungi drive evolution of plant synapses. Front Cell Infect Microbiol 3:1–9

Baluška F, Volkmann D, Menzel D (2005) Plant synapses: actin-based domains for cell-to-cell communication. Trends Plant Sci 10:106–111

Baluška F, Mancuso S, Volkmann D, Darwin F (2009) The 'root-brain' hypothesis of Charles and Francis Darwin. Plant Signal Behav 4:1121–1127

Baluška F, Mancuso S, Volkmann D, Barlow PW (2010) Root apex transition zone: a signalling-response nexus in the root. Trends Plant Sci 15:402–408

Barabási A-L, Albert R (1999) Emergence of scaling in random networks. Science 286:509–512

Barbey AK (2017) Network neuroscience theory of human intelligence. Trends Cogn Sci 22:8–20

Barto EK, Hilker M, Muller F, Mohney BK, Weidenhamer JD, Rillig MC (2011) The fungal fast lane: common mycorrhizal networks extend bioactive zones of allelochemicals in soils. PLoS One 6:e27195

Bascompte J (2009) Mutualistic networks. Front Ecol Environ 7:429e436

Beiler KJ, Durall DM, Simard SW, Maxwell SA, Kretzer AM (2010) Architecture of the wood-wide web: *Rhizopogon* spp. genets link multiple Douglas-fir cohorts. New Phytol 185:543–553

Beiler KJ, Simard SW, LeMay V, Durall DM (2012) Vertical partitioning between sister species of *Rhizopogon* fungi on mesic and xeric sites in an interior Douglas-fir forest. Mol Ecol 21:6163–6174

Beiler KJ, Simard SW, Durall DM (2015) Topology of tree-mycorrhizal fungus interaction networks in xeric and mesic Douglas-fir forests. J Ecol (3):616–628

Bierdrzycki ML, Jilany TA, Dudley SA, Bais HP (2010) Root exudates mediate kin recognition in plants. Commun Integr Biol 3:28–35

Bingham MA, Simard SW (2012) Ectomycorrhizal networks of old *Pseudotsuga menziesii* var. *glauca* trees facilitate establishment of conspecific seedlings under drought. Ecosystems 15:188–199

Boddy L, Jones TH (2007) Mycelial responses in heterogeneous environments: parallels with macroorganisms. In: Gadd G, Watkinson SC, Dyer P (eds) Fungi in the environment. Cambridge University Press, Cambridge, pp 112–158

Bowery NG, Smart TG (2006) GABA and glycine as neurotransmitters: a brief history. Br J Pharmacol 147:S109–S119

Bray D (2003) Molecular networks: the top-down view. Science 301:1864–1865

Brenner ED, Stahlberg R, Mancuso S, Vivanco J, Baluška F, Van Volkenburgh E (2006) Plant neurobiology: an integrated view of plant signaling. Trends Plant Sci 11:413–419

Brownlee C, Duddridge J, Malibari A, Read D (1983) The structure and function of mycelial systems of ectomycorrhizal roots with special reference to their role in forming inter-plant connections and providing pathways for assimilate and water transport. Plant Soil 71:433–443

Craik F, Bialystok E (2006) Cognition through the lifespan: mechanisms of change. Trends Cogn Sci 10:131–148

Darwin CR (1880) The power of movement in plants. John Murray, London

Dehaene S, Sergent C, Changeux J-P (2003) A neuronal network model linking subjective reports and objective physiological data during conscious perception. Proc Natl Acad Sci USA 100:8520–8525

Deslippe JR, Simard SW (2011) Below-ground carbon transfer among *Betula nana* may increase with warming in Arctic tundra. New Phytol 192:689–698

Deslippe JR, Hartmann M, Grayston SJ, Simard SW, Mohn WW (2016) Stable isotope probing implicates Cortinarius collinitus in carbon transfer through ectomycorrhizal mycelial networks in the field. New Phytol 210:383–390

Dudley SA, File AL (2008) Kin recognition in an annual plant. Biol Lett 3:435–438

Dudley SA, Murphy GP, File AL (2013) Kin recognition and competition in plants. Funct Ecol 27:898–906

Eason WR, Newman EI, Chuba PN (1991) Specificity of interplant cycling of phosphorus: the role of mycorrhizas. Plant Soil 137:267–274

Egerton-Warburton LM, Querejeta JI, Allen MF (2007) Common mycorrhizal networks provide a potential pathway for the transfer of hydraulically lifted water between plants. J Exp Bot 58:1473–1483

Ek H, Andersson S, Söderström B (1996) Carbon and nitrogen flow in silver birch and Norway spruce connected by a common mycorrhizal mycelium. Mycorrhiza 6:465–467

Ekblad A, Huss-Danell K (1995) Nitrogen fixation by *Alnus incana* and nitrogen transfer from *A. incana* to *Pinus sylvestris* influenced by macronutrient and ectomycorrhiza. New Phytol 131:453–459

Faust K, Raes J (2012) Microbial interactions: from networks to models. Nat Rev Microbiol 10:538–550

File AL, Klironomos J, Maherali H, Dudley SA (2012a) Plant kin recognition enhances abundance of symbiotic microbial partner. PLoS One 7:e45648

File AL, Murphy GP, Dudley SA (2012b) Fitness consequences of plants growing with siblings: reconciling kin selection, niche partitioning and competitive ability. Proc R Soc B 279:209–218

Finlay RD (1989) Functional aspects of phosphorus uptake and carbon translocation in incompatible ectomycorrhizal associations between *Pinus sylvestris* and *Suillus grevillei* and *Boletinus cavipes*. New Phytol 112:185–192

Finlay RD, Read DJ (1986) The structure and function of the vegetative mycelium of ectomycorrhizal plants. II. The uptake and distribution of phosphorus by mycelial strands interconnecting host plants. New Phytol 103:157–165

Gagliano M (2012) Green symphonies: a call for studies on acoustic communication in plants. Behav Ecol 24:289–796

Gagliano M (2014) In a green frame of mind: perspectives on the behavioural ecology and cognitive nature of plants. AoB Plants 7:plu075

Gagliano M, Grimonprez M (2015) Breaking the silence – language and the making of meaning in plants. Ecophys 7:145–152

Garcia-Garrido JM, Ocampo JA (2002) Regulation of the plant defence response in arbuscular mycorrhizal symbiosis. J Exp Bot 53:1377–1386

Giovannetti M, Sbrana C, Avio L, Stranil P (2004) Patterns of belowground plant interconnections established by means of arbuscular mycorrhizal networks. New Phytol 164:175–181

Giovannetti M, Avio L, Fortuna P, Pellegrino E, Sbrana C, Strani P (2005) At the root of the Wood Wide Web: self recognition and non-self incompatibility in mycorrhizal networks. Plant Signal Behav 1:1–5

Gorzelak M (2017) Kin selected signal transfer through mycorrhizal networks in Douglas-fir. PhD Dissertation. University of British Columbia, Vancouver, Canada

Gorzelak M, Asay AK, Pickles BJ, Simard SW (2015) Inter-plant communication through mycorrhizal networks mediates complex adaptive behaviour in plant communities. AoB Plants 7: plv050

Gyuricza V, Thiry Y, Wannijn J, Declerck S, de Boulois HD (2010) Radiocesium transfer between *Medicago truncatula* plants via a common mycorrhizal network. Environ Microbiol 12:2180–2189

He X-H, Critchley C, Bledsoe C (2003) Nitrogen transfer within and between plants through common mycorrhizal networks (CMNs). Crit Rev Plant Sci 22:531–567

He XH, Critchley C, Ng H, Bledsoe C (2004) Reciprocal N ($^{15}NH_4^+$ or $^{15}NO_3^-$) transfer between non-N_2-fixing *Eucalyptus maculata* and N_2-fixing *Casuarina cunninghamiana* linked by the ectomycorrhizal fungus *Pisolithus* sp. New Phytol 163:629–640

He XH, Critchley C, Ng H, Bledsoe C (2005) Nodulated N_2-fixing *Casuarina cunninghamiana* is the sink for net N transfer from non-N_2-fixing *Eucalyptus maculata* via an ectomycorrhizal fungus *Pisolithus* sp. supplied as ammonium nitrate. New Phytol 167:897–912

He XH, Bledsoe CS, Zasoski RJ, Southworth D, Horwath WR (2006) Rapid nitrogen transfer from ectomycorrhizal pines to adjacent ectomycorrhizal and arbuscular mycorrhizal plants in a California oak woodland. New Phytol 170:143–151

He XH, Xu M, Qiu GY, Zhou J (2009) Use of ^{15}N stable isotope to quantify nitrogen transfer between mycorrhizal plants. J Plant Ecol 2:107–118

Heaton L, Obara B, Grau V, Jones N, Nakagaki T, Boddy L, Fricker MD (2012) Analysis of fungal networks. Fungal Biol Rev 26:12e29

Heil M, Karban R (2009) Explaining evolution of plant communication by airborne signals. Trends Ecol Evol 25:137–144

Hobbie E, Agerer R (2010) Nitrogen isotopes in ectomycorrhizal sporocarps correspond to belowground exploration types. Plant Soil 327:71–83

Hocking MD, Reynolds JD (2011) Impacts of Salmon on riparian plant diversity. Science 331:1609–1612

Horton TR (ed) (2015) Mycorrhizal networks. Ecological studies. Netherlands: Springer, 224

Humphreys CP, Franks PJ, Rees M, Bidartondo MI, Leake JR, Beerling DJ (2010) Mutualistic mycorrhiza-like symbiosis in the most ancient group of land plants. Nat Commun 1:103

Ingham RE, Trofymow JA, Ingham ER, Coleman DC (1985) Interactions of bacteria, fungi, and their nematode grazers: effects on nutrient cycling and plant growth. Ecol Monogr 55:119–140

Karban R, Shiojiri K (2009) Self-recognition affects plant communication and defense. Ecol Lett 12:502–506

Karban R, Yang LH, Edwards KF (2014) Volatile communication between plants that affects herbivory: a meta-analysis. Ecol Lett 17:44–52

Karst J, Erbilgin N, Pec GJ, Cigan PW, Najar A, Simard SW, Cahill JF Jr (2015) Ectomycorrhizal fungi mediate indirect effects of a bark beetle outbreak on secondary chemistry and establishment of pine seedlings. New Phytol 208:904–914

Kretzer AM, Luoma DL, Molina R, Spatafora JW (2003) Taxonomy of the *Rhizopogon vinicolor* species complex based on analysis of ITS sequences and microsatellite loci. Mycologia 95:480–487

Kurz WA, Dymond CC, Stinson G, Rampley GJ, Neilson ET, Carroll AL, Ebata T, Safranyik L (2008) Mountain pine beetle and forest carbon feedback to climate change. Nature 452:987–990

Leake J, Johnson D, Donnelly D, Muckle G, Boddy L, Read D (2004) Networks of power and influence: the role of mycorrhizal mycelium in controlling plant communities and agroecosystem functioning. Can J Bot 82:1016–1045

Lehto T, Zwiazek JJ (2011) Ectomycorrhizas and water relations of trees: a review. Mycorrhiza 21:71–90
Levin SA (2005) Self-organization and the emergence of complexity in ecological systems. Bioscience 55:1075
Lian C, Narimatsu M, Nara K, Hogetsu T (2006) *Tricholoma matsutake* in a natural *Pinus densiflora* forest: correspondence between above- and below- ground genets, association with multiple host trees and alteration of existing ectomycorrhizal communities. New Phytol 171:825–836
Lilleskov EA, Hobbie EA, Horton TR (2011) Conservation of ectomycorrhizal fungi: exploring the linkages between functional and taxonomic responses to anthropogenic N deposition. Fungal Ecol 4:174–183
Margulis L (1981) Symbiosis in cell evolution. WH Freeman Company, San Francisco
Martin F, Stewart GR, Genetet I, Le Tacon F (1986) Assimilation of $^{15}NH_4$ by beech (*Fagus sylvatica* L.) ectomycorrhizas. New Phytol 102:85–94
McNickle GG, St. Clair CC, Cahill JF Jr (2009) Focusing the metaphor: plant root foraging behavior. Trends Ecol Evol 24:419–426
Meding SM, Zasoski RJ (2008) Hyphal-mediated transfer of nitrate, arsenic, cesium, rubidium, and strontium between arbuscular mycorrhizal forbs and grasses from a California oak woodland. Soil Biol Biochem 40:126–134
Molina R (2013) Rhizopogon. In: Cairney JWG, Chamber SM (eds) Ectomycorrhizal fungi: key Genera in profile. Springer Verlag, Berlin, pp 129–152
Molina R, Horton TR (2015) Mycorrhiza specificity: its role in the development and function of common mycelial networks. In: Horton TR (ed) Mycorrhizal networks, Ecological studies, vol 224. Springer, Netherlands, pp 1–39
Molina R, Massicotte H, Trappe J (1992) Specificity phenomena in mycorrhizal symbioses: community-ecological consequences and practical implications. In: Allen MF (ed) Mycorrhizal functioning: an integrative plant–fungal process. Chapman and Hall, New York, pp 357–423
Nehls U, Grunze B, Willmann M, Reich M, Küster H (2007) Sugar for my honey: carbohydrate partitioning in ectomycorrhizal symbiosis. Phytochemistry 68:82–91
Novoplansky A (2009) Picking battles wisely: plant behaviour under competition. Plant Cell Environ 32:726–741
Pelagio-Flores R, Ortíz-Castro R, Méndez-Bravo A, Macías-Rodríguez L, López-Bucio J (2011) Serotonin, a tryptophan-derived signal conserved in plants and animals, regulates root system architecture probably acting as a natural auxin inhibitor in *Arabidopsis thaliana*. Plant Cell Physiol 52:490–508
Perry DA, Margolis H, Choquette C, Molina R, Trappe JM (1989) Ectomycorrhizal mediation of competition between coniferous tree species. New Phytol 112:501–511
Pickles BJ, Wilhelm R, Asay AK, Hahn A, Simard SW, Mohn WW (2016) Transfer of ^{13}C between paired Douglas-fir seedlings reveals plant kinship effects and uptake of exudates by ectomycorrhizas. New Phytol 214:400–411
Poorter H, Niklas KJ, Reich PB, Oleksy J, Poot P, Mommer L (2012) Biomass allocation to leaves, stems and roots: meta-analyses of interspecific variation and environmental control. New Phytol 193:30–50
Pozo MJ, López-Ráez JA, Azcón-Aguilar C, García-Garrido JM (2015) Phytohormones as integrators of environmental signals in the regulation of mycorrhizal symbioses. New Phytol 205:1431–1436
Querejeta JI, Egerton-Warburton LM, Prieto I, Vargas R, Allen MF (2012) Changes in soil hyphal abundance and viability can alter the patterns of hydraulic redistribution by plant roots. Plant Soil 355:63–73
Rai V (2002) Role of amino acids in plant responses to stresses. Biol Plant 45:481–487
Rygiewicz PT, Anderson CP (1994) Mycorrhizae alter quality and quantity of carbon allocated below ground. Nature 369:58–60

Scheffer M, Carpenter S, Foley JA, Folke C, Walker B (2001) Catastrophic shifts in ecosystems. Nat Rev 413:591–596

Selosse M-A, Richard F, He X, Simard SW (2006) Mycorrhizal network: des liaisons dangereuses? Trends Ecol Evol 21:621–628

Semchenko M, John EA, Hutchings MJ (2007) Effects of physical connection and genetic identity of neighbouring ramets on root-placement patterns in two clonal species. New Phytol 176:644–654

Simard SW (2012) Mycorrhizal networks and seedling establishment in Douglas-fir forests (Chapter 4). In: Southworth D (ed) Biocomplexity of plant–fungal interactions, 1st edn. Wiley, Chichester, pp 85–107. isbn-10:0813815940 | isbn-13:978-0813815947

Simard SW, Durall DM (2004) Mycorrhizal networks: a review of their extent, function, and importance. Can J Bot 82:1140–1165

Simard SW, Beiler KJ, Bingham MA, Deslippe JR, Philip LJ, Teste FP (2012) Mycorrhizal networks: mechanisms, ecology and modelling. Fungal Biol Rev 26:39–60

Simard SW, Martin K, Vyse A, Larson B (2013) Meta-networks of fungi, fauna and flora as agents of complex adaptive systems. In: Puettmann K, Messier C, Coates KD (eds) Managing world forests as complex adaptive systems: building resilience to the challenge of global change, vol 7. Routledge, New York, pp 133–164

Simard SW, Asay AK, Beiler KJ, Bingham MA, Deslippe JR, He X, Philip LJ, Song Y, Teste FP (2015) Resource transfer between plants through ectomycorrhizal networks. In: Horton TR (ed) Mycorrhizal networks, Ecological studies, vol 224. Springer, Netherlands, pp 133–176

Smith S, Read D (2008) Mycorrhizal symbiosis. Academic, London

Smith SE, Smith FA (1990) Structure and function of the interfaces in biotrophic symbioses as they relate to nutrient transport. New Phytol 114:1–38

Song YY, Zeng RS, Xu JF, Li J, Shen X, Yihdego WG (2010) Interplant communication of tomato plants through underground common mycorrhizal networks. PLoS One 5:e13324

Song YY, Ye M, Li C, He X, Zhu-Salzman K, Wang RL, Su YJ, Luo SM, Zheng RS (2014) Hijacking common mycorrhizal networks for herbivore-induced defence signal transfer between tomato plants. Sci Rep 4:3915

Song YY, Simard SW, Carroll A, Mohn WW, Zheng RS (2015) Defoliation of interior Douglas-fir elicits carbon transfer and defense signalling to ponderosa pine neighbors through ectomycorrhizal networks. Sci Rep 5:8495

Southworth D, He X-H, Swenson W, Bledsoe CS (2005) Application of network theory to potential mycorrhizal networks. Mycorrhiza 15:589–595

Taylor AFS, Gebauer G, Read DJ (2004) Uptake of nitrogen and carbon from double-labelled (^{15}N and ^{13}C) glycine by mycorrhizal pine seedlings. New Phytol 164:383–388

Teste FP, Simard SW, Durall DM, Guy RD, Jones MD (2009) Access to mycorrhizal networks and roots of trees: importance for seedling survival and resource transfer. Ecology 90:2808–2822

Teste FP, Simard SW, Durall DM, Guy RD, Berch SM (2010) Net carbon transfer between *Pseudotsuga menziesii* var. *glauca* seedlings in the field is influenced by soil disturbance. J Ecol 98:429–439

Toju H, Sato H, Tanabe AS (2014) Diversity and spatial structure of belowground plant– fungal symbiosis in a mixed subtropical forest of ectomycorrhizal and arbuscular mycorrhizal plants. PLoS One:e86566

Trappe JM (1987) Phylogenetic and ecologic aspects of mycotrophy in the angiosperms from an evolutionary standpoint. In: Safir GR (ed) Ecophysiology of VA mycorrhizal plants. CRC Press, Florida

Treu R, Karst J, Randall M, Pec GJ, Cigan P, Simard SW, Cooke J, Erbilgin N, Cahill JF Jr (2014) Decline of ectomycorrhizal fungi following mountain pine beetle infestation. Ecology 95:1096–1103

Twieg B, Durall DM, Simard SW (2007) Ectomycorrhizal fungal succession in mixed temperate forests. New Phytol 176:437–447

Van der Heijden MGA, Hartmann M (2016) Networking in the plant microbiome. PLoS Biol 14: e1002378

Van der Heijden MGA, Horton TR (2009) Socialism in soil? The importance of mycorrhizal fungal networks for facilitation in natural ecosystems. J Ecol 97:1139–1150

Van Dorp C (2016) *Rhizopogon* mycorrhizal networks with interior douglas-fir in selectively harvested and non-harvested forests. Master of Science Thesis, University of British Columbia

Wipf D, Ludewig U, Tegeder M, Rentsch D, Koch W, Frommer WB (2002) Conservation of amino acid transporters in fungi, plants and animals. Trends Biochem Sci 27:139–147

Yang H, Bognor M, Steinhoff Y-D, Ludewig U (2010) H^+-independent glutamine transport in plant root tips. PLoS One 5:e8917

Inside the Vegetal Mind: On the Cognitive Abilities of Plants

Monica Gagliano

> *We can never see beyond a choice we don't understand*
> The Matrix Reloaded (2003)

Abstract Across all species, individuals thrive in complex ecological systems, which they rarely have complete knowledge of. To cope with this uncertainty and still make good choices while avoiding costly errors, organisms have developed the ability to exploit key features associated with their environment. That through experience, humans and other animals are quick at learning to associate specific cues with particular places, events and circumstances has long been known; the idea that plants are also capable of learning by association had never been proven until recently. These recent findings that experimentally demonstrated associative learning in plants not only qualify them as proper subjects of cognitive research, but in so doing, they officially open the door for the empirical exploration of cognitive processes like learning, decision-making and awareness in plants. This brief closing chapter considers the wider implications of this research by concluding that the current fundamental premise in cognitive science—that we must understand the precise neural underpinning of a given cognitive feature in order to understand the evolution of cognition and behaviour—needs to be reimagined.

An earlier version of this essay was published in the journal *Communicative & Integrative Biology* (DOI: 10.1080/19420889.2017.1288333).

M. Gagliano (✉)
Sydney Environment Institute, University of Sydney, Sydney, NSW, Australia

Centre for Evolutionary Biology, School of Biological Science, University of Western Australia, Perth, WA, Australia
e-mail: monica.gagliano@uwa.edu.au

1 Introduction

Big and small myriads of choices are made every day about everything. Based on past experiences and shaped by one's preferences, motivations and the expectations of where they may lead, choices are made through the process of decision-making, a way of pruning out presumably bad options in order to select the best ones possible. Of course, not all choices are always "the best" possible, and suboptimal (maladaptive) decision-making behaviours have been observed in both humans and non-human animals (e.g. gambling behaviour, Zentall and Stagner 2011). Poor choices are likely to affect performance and survival in many biological systems (including human societies). Therefore, individuals have evolved a remarkable capacity for making overall good decisions to successfully achieve their ends. This capacity to make sound decisions is not simply hard-wired in a behavioural blueprint but is a learned skill that can be developed and honed through sensory experiences. This implies that the complex computational processing that enables the faculty of decision-making is closely reliant on internal representations of one's historical experience (i.e. memory), developed and stored over the course of the learning process. By remembering what happened when (i.e. recollecting the past) as well as what to do when (i.e. anticipating the future), these representations inform on what is not in the immediate environment, thereby "extending" the amount of information available to the perceptual system in the present. In humans, such "extension of perception" that allows an individual to infer possible causal relationships and evaluate what opportunities are "afforded" by a given environment (i.e. "affordances" à la Gibson 1977) is defined as thinking (Hastie and Dawes 2010). This core capacity of simulating or representing information of absent objects and using the information in flexible ways in order to predict or anticipate an external event and align behaviour to the current state of the world has also become increasingly evident in several non-human animal species (Shettleworth 2001; Pickens and Holland 2004; Cheke and Clayton 2010; Crystal 2013). Many of the examples come from experiments conducted in the context of associative learning, where even "simple" conditioning tasks can result in complex representations and the behavioural flexibility generally attributed to "higher" learning (Shettleworth 2001; Andrews 2015).

While the range of complexity of these representations may remain an ongoing point of discussion, the fact is that classical conditioning in both human and non-human animals has provided a powerful framework for exploring processes like learning, memory, anticipation, awareness, decision-making and more, which are, broadly speaking, attributes of what we call the mind. Recently, this classical conditioning approach has been successfully applied to the vegetal world (Gagliano et al. 2016). Using a Y-maze task, our latest study demonstrated that seedlings of the garden pea (*Pisum sativum*) are able to acquire learned associations to guide their foraging behaviour and ensure survival. The ability of seedlings to anticipate both the imminent arrival of light ("when") and its direction ("where") based on the presence and position of a neutral conditioning stimulus (CS) demonstrates that plants are able to encode both temporal and spatial information and modify their behaviour flexibly. By revealing that plants, too, are capable of associative learning and, consequently, qualified as proper subjects of cognitive research (as discussed previously, Gagliano 2015), these findings invite us to earnestly think about the question of the vegetal mind.

2 The Ecology of Associative Learning: A Case for Plants

In the real world (outside the laboratories), individuals continuously encounter circumstances and events, the consequences of which are, more often than not, uncertain. The presence of uncertainty is an indication that the individual is yet to acquire the specific internal representation that reliably predicts its current environment. In other words, it needs to learn about the stimuli that are associated to and predict the occurrence of important outcomes, so that they can be anticipated with the least amount of uncertainty and, consequently, risk. This is best illustrated in the context of predator-prey interactions, where a close match between perception and actual reality is advantageous as it allows individuals to avoid mistakes that could have fatal consequences. In a wide range of animal species, for example, naïve prey individuals can learn to recognize a predator by being simultaneously exposed to the cue of an injured conspecific paired with predator odour (Ferrari et al. 2010); through repetition, a prey increases its certainty associated with correctly labelling a newly learned species as a predator (Ferrari et al. 2012). Beyond the realm of predation, a plethora of animal studies across functional domains has shown that associative learning enables animals to forage efficiently while avoiding potentially poisonous food (Shettleworth 2010) and to navigate their environment (Menzel et al. 2005), securing territories (Hollis et al. 1995) and reproductive success (Dukas and Duan 2000; Adkins-Regan and MacKillop 2003; Ejima et al. 2005), highlighting the importance of associative learning in shaping adaptive behaviour in a wide range of contexts and species, including humans (Heyes 2012). Then, what about plants?

Plants have been omitted from the conversation because no experimental evidence for their ability to learn by association was available, until now (Gagliano et al. 2016). It is logical then that the adaptive value of associative learning in vegetal species has never been considered. However in light of the new evidence, it is equally logical to expect that, in plants too, associative learning has a range of ecological purposes from foraging to danger avoidance to social interactions above and below ground. For example, a 2014 study demonstrated that plants are able to detect and distinguish between the sounds of feeding caterpillars and those caused by wind or singing insects (Appel and Cocroft 2014). Like in the animal example provided above, naïve plants could learn to recognize the presence of a predator by being exposed to the cue of an injured conspecific (i.e. volatile emission) paired with predator sound (i.e. feeding caterpillar). This herbivore-generated acoustic cue alone could then be used by the plant to mount up their chemical defences in response to subsequent threats of herbivory (Appel and Cocroft 2014). A testable hypothesis could be formulated that acoustic cues by virtue of their rapid transmission in the environment reinforce the effectiveness of other known warning signalling systems relying, for instance, on airborne volatiles (Gagliano and Renton 2013). If true, learning to associate sounds produced by feeding insects with the release of volatile emissions by plants under attack could reduce a plant's perceptual uncertainty and enable a rapid pre-emptive response to looming attacks as and when required. As we all know well, it is more effective to anticipate rather than wait for events to present themselves, especially when involving dangerous interactions. Accordingly, selection should favour associative learning mechanisms that enable plants to distinguish whether and when cues are indicative of impending harmful or attractive conditions and, thus, allow them to take advantage of new resources and avoid novel threats.

3 Not *What* but *Who* Is Learning Inside the Maze?

The ability to learn through the formation of associations involves the ability to detect, discriminate and categorize cues according to a dynamic *internal value system* (Ginsburg and Jablonka 2007a, b). This is a subjective system of *feelings* and *experiences*, motivated by the overall sensory state of the individual in the present and its extension via internal representations of the world experienced in the past, representations that, as mentioned earlier, play a fundamental role in the decision-making process by providing a reference for the kind of expectations that the individual projects in the future. By demonstrating that plants are able to learn by forming associations, our recent findings make some important insinuations in regard to vegetal subjectivity and awareness.

Firstly and for the simple reason that feelings account for the integration of behaviour and have long been recognized as critical agents of selection (Packard and Delafield-Butt 2014), plants too must evaluate their world *subjectively* and use their own experiences and feelings as functional states that motivate their choices. Through careful experimental design, the seedlings in our study were *allowed* to display a number of behavioural responses to inform us—the human observers—of their cognitive states. In our second experiment, for example, in which some seedlings were asked to make a decision on their growth direction during the evening hours when light would not normally be available, the young plants informed us of their lack of confidence in the neutral conditioning stimulus (CS) as a reliable predictor for the light—a decision they, otherwise, rendered most readily during daylight hours. The expression of the conditioned response (CR) in our study is certainly a good indicator of learning, but the absence of the CR in some experimental groups does not necessarily indicate that learning did not occur. Keeping in mind that in a conditioning experiment, what an individual does is not the same of what it knows (Bouton and Moody 2004), is it possible that those seedlings asked to perform outside the daylight hours *chose* to opt out of performing the conditioned light-foraging behaviour? Because the consequences of performing the CR at a time that is misaligned with the internal circadian signals are uncertain (but likely to be energetically costly), is it possible that the plants opted for what they perceived as (but, in actuality, was not) a "sure bet"—namely, their innate positive tropism to light—as a solution to the uncertainty problem? As a matter of fact, we do not know whether this is possible or else, but we now have an experimental framework to find out.

Secondly, the ability to have experiences and feelings, rather than mere sensations, can be explored as a facet of the ability to learn through the formation of associations. If we agree that associative learning and internal value systems based on feelings are evolutionarily linked and constitute what we may refer to as (basic) consciousness (Ginsburg and Jablonka 2007b), then plants could open a new interface into exploring the processes that have led to the emergence of consciousness (assuming it to be a process that actually emerged or a trait that was acquired through an evolutionary event). By uprooting it from the idea that it is a process or trait that occurs as the intrinsic operation of neural circuitry (Edelman and Tononi 2000; Koch 2004; Seth et al. 2006; Edelman et al. 2011) and thus it is *generated* by neurological substrates (as mentioned in the 2012 Cambridge Declaration on Animal Consciousness), plants help us to unnerve our premise that consciousness entails attributes derived from specific physical structures such as brains and neurons. Moreover, they encourage us to put forward a quantitative theory of consciousness that accounts, more adequately, for its *expression* through the incredible diversity of living species. Just what kind of theory this may be is an open question (but see integrated information theory, Tononi and Koch 2015), but at the very least, it should include analyses of behaviour and ecophysiology in a wider range of species that transcends the animal kingdom (including the human).

And lastly, questions about the cognitive capacities of animals and, specifically, animal consciousness often play a role in discussions about animal welfare and moral status. This debate has been recently extended to include plants (Marder 2013; Pelizzon and Gagliano 2015; Marder and Gagliano 2016), and as experimental evidence for the cognitive capacities of plants accrues, the controversial (or even taboo) topic regarding their welfare and moral standing and our ethical responsibility towards them can no longer be ignored.

References

Adkins-Regan E, MacKillop EA (2003) Japanese quail (*Coturnix japonica*) inseminations are more likely to fertilize eggs in a context predicting mating opportunities. Proc R Soc Lond B 270:1685–1689

Andrews K (2015) The animal mind: an introduction to the philosophy of animal cognition. Routledge, Abingdon

Appel HM, Cocroft RB (2014) Plants respond to leaf vibrations caused by insect herbivore chewing. Oecologia 175:1257–1266

Bouton ME, Moody EW (2004) Memory processes in classical conditioning. Neurosci Biobehav Rev 28:663–674

Cheke LG, Clayton N (2010) Mental time travel in animals. WIREs Cognit Sci 1:915–930

Crystal JD (2013) Remembering the past and planning for the future in rats. Behav Process 93:39–49
Dukas R, Duan JJ (2000) Potential fitness consequences of associative learning in a parasitoid wasp. Behav Ecol 11:536–543
Edelman GM, Tononi G (2000) The universe of consciousness: how matter becomes imagination. Basic Books, New York
Edelman GM, Gally JA, Baars BJ (2011) Biology of consciousness. Front Psychol 2:4
Ejima A, Smith BP, Lucas C, Levine JD, Griffith LC (2005) Sequential learning of pheromonal cues modulates memory consolidation in trainer-specific associative courtship conditioning. Curr Biol 15:194–206
Ferrari MCO, Wisenden BD, Chivers DP (2010) Chemical ecology of predator–prey interactions in aquatic ecosystems: a review and prospectus. Can J Zool 88:698–724
Ferrari MCO, Vrtelová J, Brown GE, Chivers DP (2012) Understanding the role of uncertainty on learning and retention of predator information. Anim Cogn 15:807–813
Gagliano M (2015) In a green frame of mind: perspectives on the behavioural ecology and cognitive nature of plants. AoB Plants 7:plu075
Gagliano M, Renton M (2013) Love thy neighbour: facilitation through an alternative signalling modality in plants. BMC Ecol 13:19
Gagliano M, Vyazovskiy VV, Borbely AA, Grimonprez M, Depczynski M (2016) Learning by association in plants. Sci Rep 6:38427
Gibson JJ (1977) The theory of affordances. In: Shaw R, Bransford J (eds) Perceiving, acting, and knowing: toward an ecological psychology. Erlbaum, Hillsdale, NJ, pp 67–82
Ginsburg S, Jablonka E (2007a) The transition to experiencing: I. Limited learning and limited experiencing. Biol Theory 2:218–230
Ginsburg S, Jablonka E (2007b) The transition to experiencing: II. The evolution of associative learning based on feelings. Biol Theory 2:231–243
Hastie R, Dawes RM (2010) Rational choice in an uncertain world: the psychology of judgment and decision-making. Sage, London
Heyes C (2012) Simple minds: a qualified defence of associative learning. Phil Trans R Soc B 367:2695–2703
Hollis KL, Dumas MJ, Singh P, Fackelman P (1995) Pavlovian conditioning of aggressive behavior in blue gourami fish (*Trichogaster trichopterus*): winners become winners and losers stay losers. J Comp Psychol 109:123–133
Koch C (2004) The quest for consciousness: a neurobiological approach. Roberts, Englewood, CO
Marder M (2013) Plant-thinking: a philosophy of vegetal life. Columbia University Press, New York
Marder M, Gagliano M (2016) Revolution. In: Marder M (ed) Grafts – writings on plants. University Press, Minnesota, pp 79–82
Menzel R, Greggers U, Smith A, Berger S, Brandt R, Brunke S, Bundrock G, Hülse S, Plümpe T, Schaupp F, Schüttler E, Stach S, Stindt J, Stollhoff N, Watzl S (2005) Honey bees navigate according to a map-like spatial memory. Proc Natl Acad Sci USA 102:3040–3045
Packard A, Delafield-Butt JT (2014) Feelings as agents of selection: putting Charles Darwin back into (extended neo-) Darwinism. Biol J Linn Soc 112:332–353
Pelizzon A, Gagliano M (2015) The sentience of plants: animal rights and rights of nature intersecting? AAPLJ 1:15–14
Pickens CL, Holland PC (2004) Conditioning and cognition. Neurosci Biobehav Rev 28:651–661
Seth AK, Izhikevich E, Reeke GN, Edelman GM (2006) Theories and measures of consciousness: an extended framework. Proc Natl Acad Sci USA 103:10799–10804
Shettleworth SJ (2001) Animal cognition and animal behaviour. Anim Behav 61:277–286
Shettleworth SJ (2010) Cognition, evolution and behavior. Oxford University Press, Oxford
Tononi G, Koch C (2015) Consciousness: here, there and everywhere? Phil Trans R Soc B 370:20140167
Zentall TR, Stagner J (2011) Maladaptive choice behaviour by pigeons: an animal analogue and possible mechanism for gambling (sub-optimal human decision-making behaviour). Proc R Soc Lond B 278:1203–1208

Index

A
Associative learning, 17, 18, 30, 61, 173, 216–219

B
Behavior, 36–43, 45, 46, 51–64, 118, 149, 192, 194, 195, 197, 200, 202–204, 207
Behaviourism, 19–23

C
Calcium wave, 131, 132
Chromatin assembly factor 1 (CAF-1), 84
Chromatin structure, 8, 10, 81–84, 86, 91, 92, 97, 102, 115
Cognition, 1, 9–11, 18, 20, 21, 36–38, 45, 46, 51–64, 163–165, 168–170, 173, 174, 194–195, 202, 207, 215
Cognitivism, 19, 20, 39, 169
Conditioning, 17, 19, 23, 29, 30, 41, 42, 173, 216, 218
Consciousness, 10, 11, 54, 219
Cytosine methylation, 83, 88, 94, 99, 102, 111–113, 123

D
Decreased DNA methylation 1 (DDM1), 85, 116, 117, 119–123
Dewey, J., 19, 20
Dicer-like (DCL), 100

DNA methylation, 9, 81, 83–91, 93, 94, 97, 100–102, 111–114, 133
DNA repair, 6, 81–83, 85, 88, 90–92, 99–100, 102
DNA repair factor, 91, 99–100
Double-strand break (DSB), 81, 85, 92
DSB-induced ncRNAs (diRNAs), 91, 92, 100

E
Ecological psychology, 36, 39, 40, 42, 44, 46, 171
Environment, 1, 3–10, 12, 18–22, 24–30, 39–43, 51–64, 80, 81, 83, 89, 92–95, 112, 115, 124, 125, 131–136, 147, 148, 164–175, 177, 180, 182, 183, 193, 197, 200–203, 205, 207, 216–218
Epiallele, 94, 113–118
Epigenetic factors, 81, 95, 100–101
Epigenetic inheritance, 7, 80, 95
Epigenetic recombinant inbred lines (epiRIL), 119–125
Epigenetic regulation, 80, 90, 100
Epimutation accumulation lines (epiMAs), 121–123, 125
Evolution, 3, 5, 7, 22, 25, 26, 51–53, 55, 89, 125, 131, 177, 215

F
Familiarization, 132

Index

G
Gene expression, 2, 8, 9, 83, 85–88, 97, 102, 103, 112–115, 123, 171, 172, 201, 204
Genetic variation, 111–126
Genome maintenance, 90–92
Genome stability, 79–102, 115

H
Heterochromatin decondensation, 84–85
Heterochromatin protein 1 (HP1), 83, 84
Histone acetyltransferases (HATs), 90, 91
Histone deacetylases (HDACs), 90, 91
Histone demethylases (KDMs), 90, 91
Histone modifications, 2, 4, 9, 81, 83, 90, 91, 93, 101, 102
Homologous recombination (HR), 81, 82, 86, 88, 89, 94–97, 100, 101

I
Intelligence, 17–31, 35–46, 53, 55, 163, 164, 167, 169, 193–195, 197, 207

K
KDMs, *see* Histone demethylases (KDMs)

L
Learning, 1–13, 17, 18, 20, 21, 23, 26, 28, 30, 35–46, 51, 52, 54, 55, 61, 62, 111, 112, 132, 134, 135, 140, 163–183, 191–207, 216–219

M
Maintenance of methylation 1 (MOM1), 85
Mechanistic philosophy, 18, 21
Memory, 1–13, 52, 56, 61, 93, 96, 99, 101, 103, 111–126, 132, 134, 135, 139–157, 163, 171, 172, 175–178, 180, 181, 191–208, 216
Methyl-CpG-binding domain proteins (MBDs), 84, 85
Methyl methanesulfonate (MMS), 85, 97

N
Noncoding RNAs (ncRNAs), 83, 91, 92, 97, 99, 1000–102
Non-homologous endjoining (NHEJ), 81, 82, 100

O
Oilseed rape mosaic virus (ORMV), 98

P
Plant intelligence, 17–31, 35, 36, 38–40, 42, 44–46, 53, 55
Plant learning, 35–48, 163–183
Plants, 6, 8–10, 17–31, 35–46, 51–64, 79–102, 111–126, 131–135, 139–157, 163–183, 192–195, 197, 199–204, 206, 207, 215–219
Pragmatism, 21
Pure epigenetic variation, 112, 115, 117–121, 123–125

R
Response to stress, 82–86, 89, 91–97, 99, 101, 125
RNA-direct DNA methylation (RdDM), 85, 87, 100–102, 112

S
Sensitization, 17, 132, 173
Small RNA (smRNA), 83, 86, 92
Stress memory, 96, 101

T
Teleology, 18–20
Tobacco mosaic virus (TMV), 88, 98, 99
Transactionalism, 22, 25, 28
Transgenerational inheritance, 99–101, 112, 125
Transgenerational response, 92, 94–96, 98–99, 101, 102
Transposon activity, 89

Printed by Printforce, the Netherlands